A CENTURY OF
SAND DREDGING
in the Bristol Channel
VOLUME TWO: THE WELSH COAST

PETER GOSSON

AMBERLEY

Front cover: Bowline at Redcliffe Bascule Bridge, Bristol (*c.* 1977).
Back cover: Hovering I (1984-87) on the River Usk, Newport.

First published 2011

Amberley Publishing
The Hill, Stroud
Gloucestershire GL5 4ER

www.amberley-books.com

British Library Cataloguing in Publication Data.
A catalogue record for this book is available from the British Library.

ISBN 978-1-4456-0508-1

Typesetting and Origination by FONTHILLDESIGN.
Printed in Great Britain.

Contents

TO THE MEMORY OF MY FATHER
CAPTAIN ROLAND JOHN GOSSON, MN

ALSO THE MEN OF THE BRISTOL CHANNEL
DREDGING FLEETS PAST PRESENT AND YET TO COME

MAY THE GENTLE WIND BE COOL UPON YOUR BACK
MAY THE GUIDING STAR BE CLEAR ALONG YOUR WAY
MAY THE LIGHT OF GOD SHINE ON YOU
FROM THE RISING OF THE SUN
UNTIL ITS SETTING BEYOND THE WESTERN SEA

Bowline at Redcliffe Bascule Bridge, Bristol (*c.* 1977).

Introduction

A Century of Sand Dredging in the Bristol Channel, Volume I dealt with sea-dredged aggregates and the development of sand dredgers, together with the dredging companies and their ships involved in the trade on the English side of the Bristol Channel. Volume II will deal with the developments in the sea-dredged aggregate trade, together with the companies and their ships involved on the Welsh side of the Bristol Channel.

How many times has it been said that if you have a good idea someone is sure to copy it? This was certainly true when it came to the formation of the first sand dredging companies in Wales, which were initially based in the port of Cardiff. Potential participants kept a close and careful watch on the activities and operations of the Bristol Sand & Gravel Company, the pioneers in the trade, mainly to assess if the new trade was viable and would be worthy of investment.

The first person in Cardiff to branch out into sand dredging was Arthur Sessions using his first ship, the *Duke of Edinburgh*. Later, Sessions was to become a life-long friend and trading partner of Fred Peters, the owner of Bristol Sand & Gravel Co., whose ships were to be later seen supplying the South Wales wharves of Arthur Sessions and Sons. Seeing the success of Sessions' venture in Cardiff, many small one-ship operations sprang up, some of which didn't last long enough to record as being in the trade. However, in 1920 one company set up who were later to dominate the South Wales sand trade. This company was F. Bowles and Sons. In 1932 they were instrumental with Fred Peters in setting up a management company to manage the affairs of dredger owners operating in the Bristol Channel, and it was said that this new venture took the title British Dredging because companies from both England and Wales were involved.

The Cardiff-owned ships, together with those of Bristol Sand & Gravel, had by the late 1920s started trading down channel to Swansea. Seeing this trade building up at Swansea, a local family named Bevan set up in 1932 as South Wales Sand & Gravel Company. With the establishment of this company, all the South Wales ports could be served on a regular basis by dedicated sand dredging companies. Licenses were applied for and granted to allow dredging in the Bristol Channel, and for every ton of aggregate dredged and landed, a royalty was paid to the crown. The Cardiff dredgers, like those of the Bristol fleets, dredged in the upper Bristol Channel and near the Holms, whereas the Swansea ships tended to dredge at the Nash sands, and later they were to pioneer dredging at the Helwick shoal; this gave them a great advantage in the time saved steaming to and from the dredging site.

In many ways the Welsh sand trade reflected the operations at Bristol. The early ships in the trade were elderly steam coasting vessels which were near or at the end of their coasting life and purchased, in some cases, for only a few of hundred pounds, making their purchase and conversion a very attractive proposition. It took a long time before any sort of modernisation took place, but in 1939, F. Bowles and Sons took delivery of a motor ship which had mechanical problems; because her engine was German-built, it was considered irreparable due to problems obtaining parts. This ship was the collier *Rookwood*, which Bowles managed to repair and convert for sand dredging, making her the first motor dredger in the Bristol Channel. The South Wales cities, like Bristol, suffered extensive damage from wartime bombing, and post-war redevelopment required vast quantities of aggregate, which in turn led the dredger owners to expand their fleets to cope with this demand. At Swansea in 1948, a new company, Channel Sand and Ballast, was set up; in the years immediately following the war, it had difficulty in finding a suitable ship. The solution was found in the form of an ex-government tank landing craft; after conversion in Holland, this vessel became the *Sand Moor*, which was probably the most interesting dredger conversions of all times. With the building of Bristol Sand & Gravel's *Camerton* in 1950, to supply the needs of Bristol's redevelopment, the company were able to almost permanently base both the *Saltom* and *Dunkerton* in the South Wales ports, running for Sessions and Sons. Further expansion at Cardiff saw *Bowstar*, another motor ship, join the Bowles fleet, purchased part-built in 1950 and completed as a sand dredger. It was in 1953 that the first purpose-built motor sand dredger arrived from her Dutch builders, she was the *Bowline*, and her success led both Bowles and other owners to order Dutch-built tonnage.

In 1962, a merger between F. Bowles and Sons and Bristol Sand & Gravel led to the forming of a new company, reactivating the title British Dredging Company. With further mergers, this company was to become the largest aggregate dredging company in Europe.   The new company, through Bowles and Sons, owned dry docks and ship repair facilities at Cardiff, but their expansion plans required more new ships.

British Dredging Co. purchased the controlling interest in Ailsa Shipbuilding Company at Troon in Scotland. This now meant British Dredging could not only repair their own ships, but could also build their own new replacement tonnage. It was also in 1962 that probably the strangest sand dredger arrived at Cardiff to start work for the newly formed Sand and Gravel Marketing Company. She was the ex-Mersey mud dredger *Hoyle*, which had been purchased and given the name *Sand Galore*. This may have been wishful thinking as this vessel proved to be a failure, mainly due to her size and the fact that she had no open cargo hold and could only pump discharge wet sand.

The early to mid 1960s saw the coal-fired steam ships being phased out due to the difficulties in obtaining good-quality steam coal. The last one to go was the *Dunkerton*, which was scrapped in 1965. The last steamship to survive was the oil-fired *Camerton*, but rising fuel costs and oil shortages led to her sale to Greek owners in 1973. A minor revolution in the industry took place in 1966 with the delivery of the *Hoveringham 1* from her builders at Appledore. This vessel was able to self-discharge using a bucket scraper system, which meant that in ports of call no shore cranes were required. Again, the success of this ship led to further orders for this discharge system, the down side being that it could only be fitted to new purpose-built tonnage. This development was to revolutionise the sand trade and its ships by cutting out wasteful down time during discharge. Instead of discharging in port between tides by shore crane, taking up to 12 hours, ships could now arrive on the first of the tide, discharge in 3 hours or less, and sail on the same tide. Truly, the modern sand dredger, crammed with electronics for finding and loading aggregates, with bridge control for both loading and discharge, can only be described as a floating sand factory. With three main companies serving the Bristol Channel, Hanson Aggregates Marine, Cemex UK Marine and United Marine Aggregates, it is good to see that there is still one independent operator at Newport in the form of Severn Sands. The aggregate industry of today is a far cry from the industry founded by the pioneers Bristol Sand & Gravel in 1912.

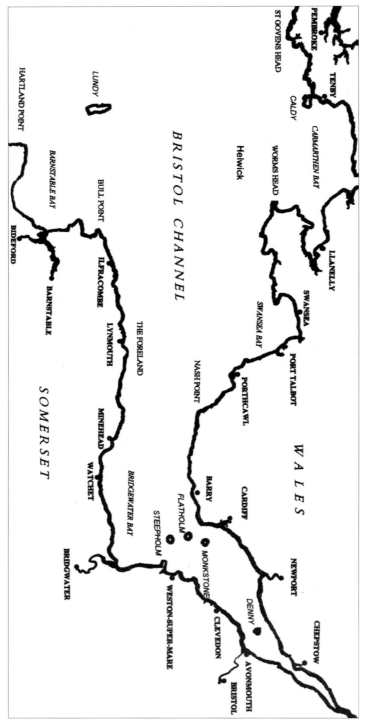

A sketch of the Bristol Channel (not to scale and all positions are approximate).

CHAPTER 1

The Companies Trading and Their Development

Cemex UK Marine Ltd

The origins of Cemex UK Marine date to 1962 and the formation of British Dredging by the merger of Bristol Sand & Gravel and F. Bowles & Sons of Cardiff. The company formed from this merger became one of the largest in the aggregate industry, thanks to massive financial investments and the takeover of rival aggregate dredging firms in the Bristol Channel, including Arthur Sessions & Son, which had earlier purchased the dredgers of William Holway and George Binding. With the purchase of Western Dredgers of Newport came the interests of J. & R. Griffiths, Newport Sand & Gravel, and Swansea Sand & Gravel. With these acquisitions, the company expanded its UK markets and also moved into Europe.

Having sold their joint venture interest in the Bristol sand trade to ARC Marine, British Dredging Aggregates of Cardiff announced they were to open a new sand wharf at Avonmouth Dock; this wharf, at M Berth in Avonmouth Old Dock, opened on 5 November 1992.

British Dredging were by this time in a further joint venture. This time their partner was Ready Mix Concrete, which had been operating in the Bristol Channel since 1985. Ready Mix Concrete was based on the south coast, their headquarters at Southampton. Its move into the Bristol Channel came when it

RMC Marine Ltd and Cemex UK Marine company logos.

bought out Bury Sand in Swansea. The agreement with British Dredging was for them to operate the Bristol Channel business.

The origins of Ready Mix Concrete date to 1947. In 1956, rapid expansion took place and a new company name, South Coast Shipping Ltd, was adopted. Later, Ready Mix Concrete (or RMC Ltd) gained full control of South Coast Shipping and became RMC Marine.

In 2005, the group was bought out by Cemex of Mexico; it now operates as Cemex UK Marine Ltd.

Arthur Sessions & Son

Arthur Sessions & Son of Cardiff was already established as a builders' merchants by the time it went into ship owning in the latter part of 1911. The 1912 *Lloyd's Register of Shipping* shows it had two business addresses; the established business was at 53 Penarth Road, Cardiff, and a secondary address was at Gloucester Docks. The early vessels were engaged in the short sea coastal trade to and from Cardiff and Gloucester; they were also engaged in the carriage of builders' supplies and sanitary chinaware. At the head office in Penarth Road, Sessions & Son had a large warehouse and showroom. From here, all types of builders' materials and sanitary chinaware could be purchased by the building trade.

By 1914, Arthur Sessions' first ship, the *Duke of Edinburgh*, is known to have been involved in the recovery of aggregate from the Bristol Channel. The crew of these early vessels excavated sand by manually digging and loading the cargo; suction dredging was established in Cardiff in around 1919/20. These early Cardiff sand vessels ran cargos of sand and other coastal cargos. When the sand trade proved to be successful, coastal trading ceased.

Arthur Sessions & Son established a wharf for their sand trade in the Glamorganshire Canal at Cardiff. By 1921, Welsh Sand & Gravel, George Hamlin (Sandridge & Co.), F. Bowles & Sons, and J. & R. Griffiths were also working in the sand trade out of the canal. The closure of the canal led Sessions & Son to find alternative wharfage at Cardiff. The company moved and set up two wharves in the Bute West Dock – one to handle sand, the other to handle gravel cargos.

Through the years that followed, customer demand saw Arthur Sessions set up new sand and gravel wharves at Newport, in the River Usk, and at the South Dock at Swansea. Each depot was capable of storing aggregate and had its own fleet of lorries for distribution throughout South Wales. A trading agreement was formed between Arthur Sessions & Son and Bristol Sand & Gravel for the latter's ships to trade with, and supply, the South Wales wharves.

Arthur Sessions owned and managed his own ships from his beginnings in ship owning in 1911 until 1932, when he joined the newly formed British Dredging Co., a management company run by Henry Sockette.

The post-war rebuilding of South Wales placed heavy demands on the aggregate suppliers, but unlike most of the other owners, Arthur Sessions chose not to build new tonnage as a result of increased business. In 1952, he sold two of his dredgers for scrap, and continued to trade with one ship, depending on Bristol Sand & Gravel to supply his wharves in South Wales. In 1961 he sold his last ship, the *Endcliffe*, to John Cashmore at Newport for breaking up. Until 1960, the company had a warehouse at the Gloucester Docks.

In 1962, with the merger of Bristol Sand & Gravel with F. Bowles – forming British Dredging – Arthur Sessions & Son found themselves under pressure from the new company, who, in order to expand, were seeking further mergers. A turning point was reached in 1965, when Arthur Sessions & Son sold their interests in the sand trade to British Dredging. John Sessions, the son of the founder, became sand sales manager for British Dredging, and soon afterwards all traces of Sessions & Son disappeared within British Dredging.

Fleet List – Arthur Sessions & Son

DUKE OF EDINBURGH (no image available)
In fleet as a sand dredger 1914-1933
Built in 1873 by Humphrys, Pearson & Co., Hull, as the support vessel/coaster *Duke of Edinburgh*
Gross tonnage 157 – Length 114.5 feet – Beam 19.9 feet
Machinery: Steam compound two cylinder 25 rhp, by Humphrys, Pearson & Co.
History:
1873 Completed for the Corporation of Trinity House, Hull
1896 Sold to J. MacCausland of Hull
1911 Sold to Arthur Sessions of Cardiff; used in the short sea coastal trade
1914 Involved in the Cardiff sand trade
1919/20 Converted for suction dredging
1933 Sold for further trading; scrapped by her new owners

LOCH NELL (no image available)
Involved in the sand trade 1919-22
Built in 1877 by H. M. McIntyre & Co., Paisley, as the steam coaster *Loch Nell*
Gross tonnage 123 – Length 89.5 feet – Beam 17.1 feet
Machinery: Steam compound two cylinder 30 rhp, by Muir & Houston, Glasgow
History:
1877 Completed for W. Sim & Co. of Glasgow; registered in Glasgow
1917 Owned by J. G. Stewart of Glasgow
1919 Sold to Arthur Sessions & Son, Cardiff; used in the short sea coastal

trade and the Cardiff sand trade
1922 Vessel no longer listed

MANLEY (no image available)
Involved in the sand trade 1922-35
Built in 1888 by Head & Bernard, Hull, as the iron built coaster *Manley*
Gross tonnage 117 – Length 91 feet – Beam 17.2 feet
Machinery: Steam compound two cylinder 17 rhp, by Cochrane & Co., Birkenhead
History:
1888 Completed for Bennetts & Co., Grimsby; registered in Grimsby
1920 Sold to Arthur Sessions & Son; reregistered in Cardiff; used in the short sea coastal trade
1922 Converted into a sand suction dredger in Cardiff; employed in the South Wales sand trade
1935 Sold to John Cashmore; scrapped in Newport

BLOODHOUND (no image available)
In fleet as a sand dredger 1921-37
Built in 1890 by Cook, Welton & Gemmell, Hull, as the iron-built screw ketch *Bloodhound*
Gross tonnage 156 – Length 100.4 feet – Beam 20.9 feet
Machinery: Steam compound two cylinder 45 rhp, by Bailey & Leetham, Hull
History:
1890 Completed for the Humber Steam Trawling Co.; registered in Hull
1916 Sold to T. Lauder & Co., Aberdeen; reregisered in Aberdeen
1920 Sold to Oberon Shipping Co., also of Aberdeen
1921 Sold to Arthur Sessions & Son; reregistered in Cardiff, converted into a sand suction dredger; employed in the South Wales sand trade
1937 Sold to John Cashmore; scrapped in Newport

KENDY (no image available)
In fleet as a sand dredger 1922-29
Built in 1919 by J. Crichton & Co., Chester, as the steam coaster *Kendy*
Gross tonnage: 139 – Length: 93.4 feet – Beam: 18.6 feet
Machinery: Steam compound two cylinder, by Rutherford & Co.
History:
1919 Completed for Westville Shipping, Cardiff; used as a coastal cargo vessel; managed by G. H. Allen
1922 Sold to Arthur Sessions & Son; converted into a sand suction dredger
1929 Vessel not listed; presumed sold or scrapped

LYNDALE

Lyndale (1935-46).

In fleet as a sand dredger 1935-46
Build details: See William Holway
History:
1935 Purchased by Arthur Sessions & Son from William Holway as the *Lyndale*
1946 Sold to W. R. Metcalf of Ilfracombe

CORBEIL

Corbeil (1936-38) at Minehead.

In fleet as a sand dredger 1936-38
Build details: Bristol Sand & Gravel Co. (*Volume One: The English Coast*)
History:
1936 Purchased by Arthur Sessions & Son from Willoughby & Lamb as the *Corbeil*
1938 Sold to John Cashmore; scrapped in Newport

SANDALE

Sandale (1937-52) at Cardiff.

In fleet as a sand dredger 1937-52
Build details: See George Binding
History:
1937 Sold to Arthur Sessions & Son by George Binding as the *Evelyn B*; renamed *Sandale*
1952 Sold to John Cashmore; scrapped in Newport

ENDCLIFFE

Endcliffe (1939-61) near Swansea.

In fleet as a sand dredger 1939-61
Built in 1911 by W. Walker, Maryport, as the fast steam coaster *Endcliffe*
Gross tonnage 367 – Length 135.2 feet – Beam 25.1 feet
Machinery: Steam compound two cylinder, by J. Richie of Gloucester
History:
1911 Built and completed for Thomas W. Ward; registered in Liverpool; she is a very fast ship, with a large boiler
1939 Sold to Arthur Sessions & Son, Cardiff; reregistered in Cardiff
1939 Converted into a sand suction dredger; proves costly to run
1948 A smaller boiler made in 1935 is fitted; she is now very slow
1961 Sold to John Cashmore; scrapped in Newport

F. Bowles & Sons

The Cardiff-based, family-controlled company F. Bowles & Sons rose from small beginnings in the haulage business to become a major supplier of sea-dredged aggregate and building materials throughout South Wales. Their origins date from 1896, when Mr. F. Bowles started a haulage business in Cardiff with a horse and cart. The company expanded under his successors, until in 1920 it could buy ships and venture into aggregate dredging. Initially, the company ran its dredging business from a small wharf on the bank of the Glamorganshire Canal at Harrowby Street, Cardiff. It was from here that the aggregate dredged from the sandbanks in the Bristol Channel was distributed by horse and cart to customers in Cardiff.

Later, after the closure of the Glamorganshire Canal, F. Bowles, together with other dredging concerns, was forced to move and set up new wharves in and around Cardiff Dock. F. Bowles set up a new wharf, depot, and office at Avondale Road, in the Grangetown district of Cardiff. This new wharf was known locally as 'the Ferry'.

In 1932, F. Bowles & Sons became a limited liability company. At the same time they formed, with Bristol Sand & Gravel, the British Dredging Co. a non-profit management company for a pool of sand dredgers operated in the Bristol Channel by five independent owners.

The company further expanded its interests in Cardiff by moving into ship repair with the acquisition of East Bute Dry Dock, Cardiff. Here, the company repaired and converted its own ships into sand dredgers, and offered a commercial repair service to other owners. During the great depression in the 1930s, F. Bowles & Sons purchased surplus, redundant, and damaged tonnage, which they either repaired for their own use, or sold as the shipping markets picked up. The success of the venture into sand dredging led them to expand further in this field; using Cardiff as the head office, the company established aggregate wharves at Newport, Cardiff, Barry, Briton Ferry, and Swansea. All these wharves were supplied by F. Bowles & Sons ships and the aggregate was delivered on a daily basis by their fleet of lorries. As the business grew, the company was split into three divisions – distribution, shipping, and sales – which were controlled and run by members of the family.

In 1950, the company purchased a part-built motor ship from Clelands Shipbuilding Co. She was completed as a motor sand suction dredger. This ship was the *Bowstar* and was the first ship in the company fleet to carry the 'Bow' prefix. Also in early 1950, the company ventured into ready-mixed concrete and dry-pack ready mix, for which stone-crushing plants, mixing plants, and drying plants were installed at Newport and Cardiff. The stone-crushing plants were capable of crushing sea-dredged stone of up to four inches into any size or grade required, for use in the ready-mix plant.

A fleet of revolving drum lorries were purchased and operated by F. Bowles & Sons. The company could now deliver ready-mix concrete on a daily basis throughout South Wales. Major contracts were won to supply aggregate for rebuilding the cities of South Wales after the wartime bombing. The concrete was also used for large projects such as the building of the Aberthaw Power Station, and the building of the M4 through South Wales.

The company opened a land site at Morfa Mawr near Swansea, from where dune sand was extracted under licence, for use in the manufacturing of asphalt. In 1960, the company owned four modern motor dredgers, and started to look for expansion outside the Bristol Channel. The company dispatched the *Bowcrest* to look for sites around the Thames and on the north-east coast. Having found suitable dredging sites and a ready market, together with the possibility of moving into Europe at a later date, the *Bowcrest* returned to Cardiff for F. Bowles & Sons to plan its next move.

In 1962, under an agreement worked out some years before, F. Bowles & Sons and Bristol Sand & Gravel merged to form the reactivated British Dredging Co. The new company had members of the Bowles family as its directors, and soon expanded its operations to the Thames, the North East, and Europe. The new company retained the 'Bow' prefix on the names of its vessels.

Fleet List – F. Bowles & Sons

THOMOND (no image available)
In fleet as a sand dredger 1920-24
Built in 1900 by Boon Molema & De Cook, Hoogezand, as the steam coaster *Thomond*
Gross tonnage: 127
Length: 93 feet
Beam: 18.1 feet
Machinery: Steam compound two cylinder, by Molema & De Cook
History:
1900 Completed for Limerick Steamship Co.
1919 Sold to J. George of Milford; reregistered in Milford
1920 Sold to Joseph Bowles of Cardiff; used as a sand dredger
1924 Founders in heavy weather near the West Cardiff Buoy

ALEXANDRA (no image available)
In fleet as a sand dredger 1923-44
Built in 1901 by W. J. Yarwood, Northwich, as the steam coaster *Alexandra*
Gross tonnage 211 – Length 101 feet – Beam 22 feet
Machinery: Steam compound two cylinder, by W. J. Yarwood
History:
1901 Completed for the Northwich Carrying Co.
1923 Sold to Joseph Bowles; used in the Cardiff sand trade
1932 Converted from steam; fitted with new oil engine
1935 Owners become F. Bowles & Sons
1944 Sold to Hubert F. Ashmead of Bristol

DELORAINE

Deloraine (1828-56) at the Holms.

In fleet as a sand dredger 1928-56
Built in 1900 by J. McArthur & Co., Paisley, as the steam coaster *Deloraine*
Gross tonnage 265 – Length 120.2 feet – Beam 22.1 feet
Machinery: Steam compound two cylinder 37 rhp, by McLachlan & Co.
History:
1900 Completed for J. G. Few & Co. (registered as the Deloraine Steamship Co.)
1920 Sold to Heath Shipping; registered in Glasgow
1928 Sold to F. Bowles & Sons; converted into a sand suction dredger; reregistered at Cardiff; managed by Joseph Bowles; serves extensively in the South Wales sand trade
1956 Sold to John Cashmore in June; scrapped in Newport

HARTFORD

Seabourne Alfa, formerly *Hartford* (1939-50).

In fleet as a sand dredger 1939-50
Built in 1912 by J. P. Reynoldson & Sons, South Shields, as the steam coaster *Hartford*
Gross tonnage 410 – Length 144 feet – Beam 24.4 feet
Machinery: Steam compound two cylinder, by Reynoldson & Sons
History:
1912 Completed for the Northwich Carrying Co. of Liverpool
1926 Sold to W. A. Wilson of Liverpool
1929 Sold to the Cement Marketing Co. of London
1939 Sold to F. Bowles & Sons in February; converted into a sand suction dredger in the East Bute dry dock, Cardiff
1950 Sold to Seabourne Aggregates, Marchwood; renamed *Seabourne Alfa*

ROOKWOOD

Sand Martin, formerly *Rookwood* (1939-51).

In fleet as a sand dredger 1939-51
Built in 1936 by Henry Robb Ltd, Leith, as the motor collier *Rookwood*
Gross tonnage 633– Length 177 feet – Beam 28.6 feet
Machinery: Single-screw eight-cylinder diesel engine, by Humboldt Deutz
Motoren
History:
1936 Completed for William France Fenwick of London
1939 Sold to F. Bowles & Sons; reregistered in Cardiff; suffers mechanical
problems; spares for her German-built engine impossible to obtain; repaired
and converted into a sand suction dredger
1951 Sold to Zinal Steamship Co.; renamed *Sand Martin*

SUNFOLD

In fleet as a sand dredger 1948-61

Built in 1917 by Swan Hunter & Wigham Richardson, Sunderland, as the steam coaster *Northwick*

Gross tonnage 472 – Length 158.2 feet – Beam 26.1 feet

Machinery: Steam triple expansion, by W. Beardmore & Co.

History:

1917 Completed for Pile & Co. of Sunderland

1926 Sold to Coombes of Sunderland; renamed *Eskburn*

1929 Sold to W. G. James of Middlesborough

1938 Sold to British Isles Coasters

1942 Sold to Efford Shipping Co.

1947 Sold to F. Bowles & Sons

1948 Converted into a sand suction dredger renamed *Sunfold*

1961 Scrapped in the Prince of Wales dry dock, Swansea

BOWSTAR

Bowstar (1950-62) at *Swansea*.

In fleet as a sand dredger 1950-62

Built in 1950 by Clelands Successors, Wallsend; originally laid down as a motor coaster, but completed as the sand dredger *Bowstar*

Gross tonnage 561 – Length 169.4 feet – Beam 28.2 feet

Machinery: Single-screw seven-cylinder, 731 bhp diesel engine, by Ruston & Hornsby

History:

1950 Completed for F. Bowles & Sons; purchased part-built; converted into a sand dredger

1954 Her cargo suddenly shifts, causing a severe list; sails back to Newport from the Holms dredging grounds stern first, which takes over six hours; repaired and put straight back into service; trades to the South Wales ports

1962 Transfers to the British Dredging fleet

BOWLINE

Bowline (1953-62), in the River Usk, Newport.

In fleet as a sand dredger 1953-62
Built 1953 by N. V. Bodewes Schps, Martenshoek, as the purpose-built motor sand dredger *Bowline*
Gross tonnage 596 – Length 179 feet – Beam 29.2 feet
Machinery: Single-screw eight-cylinder 860 bhp diesel engine, by Masch Kiel AG.
History:
1953 Completed to the order of F. Bowles & Sons; designed on the Dutch coaster model, an improved version of the *Bowstar*; spends her working life in the Bristol Channel
1962 Transfers to the British Dredging fleet

BOWCREST

Bowcrest (1955-62) leaving Swansea.

In fleet as a sand dredger 1955-62

Built in 1955 by N. V. Bodewes Schps, Martenshoek, as the purpose-built motor sand dredger *Bowcrest*

Gross tonnage 587 – Length 179 feet – Beam 29.11 feet

Machinery: Single-screw eight-cylinder 725 bhp diesel engine, by Masch Kiel AG.

History:

1955 Completed to the order of F. Bowles & Sons, an improved version of the *Bowline*; designed on the Dutch coaster model; serves the South Wales ports

1962 Transfers to the British Dredging fleet

BOWPRIDE

Bowpride (1960-62) at the Holms.

In fleet as a sand dredger 1960-62

Built in 1960 by N. V. Bodewes Schps, Martenshoek, as the purpose-built motor sand dredger *Bowpride*

Gross tonnage 780 – Length 214.11 feet – Beam 33.6 feet

Machinery: Single-screw six-cylinder 1,250 bhp diesel engine, by Masch Kiel AG.

History:

1960 Completed to the order of F. Bowles & Sons; designed specifically with the Bristol Channel sand trade in mind

1962 Transfers to the British Dredging fleet

J. & R. Griffiths

J. & R. Griffiths was another of the smaller firms to set up in sand dredging in around 1922. Like other early sand dredging firms in Cardiff, the company operated from the Glamorganshire Canal. They remained in the canal with Sandridge & Co., running from a wharf at Harrowby Street, until the canal closed in December 1951, when they moved to a new wharf in Cardiff's West Dock.

Griffiths didn't only own ships; the company also managed ship purchases for other operators, and then managed those ships. They were instrumental in the purchase of the *Isabel* and the setting up of Newport Sand & Gravel.

In Swansea, the firm was involved in the purchase of the Imogen and the setting up of Swansea Sand & Gravel for William Adams, a local businessman. Later, under the management of J. & R. Griffiths, all three companies were merged and ran as J. & R. Griffiths of Cardiff. Prior to this, the Isabel ran mainly to Newport, but was also seen at Cardiff, and the Imogen ran to Swansea, with visits to Cardiff. Griffiths's own ship, the Indium, was worked hard in Cardiff, and also did relief work in Swansea and Newport.

Like most of the dredging companies, J. & R. Griffiths had very good times during the 1950s. But in the early 1960s, with the formation and dominance of British Dredging, small concerns like Griffiths had outlived their time. The ships were sold off, and their dredging interests were sold in 1963 to the newly formed Western Dredgers of Newport.

The company also acted as managers for the Isabel Steamship Co. and the Imogen Steamship Co.

It is understood that J. & R. Griffiths was also involved in the ownership or management of the *Skarv*, which was converted for sand dredging in the Bristol Channel in 1940. She is listed as owned by J. & R. Griffiths, whereas another listing states Davies as the owner, and Griffiths of Cardiff as the management.

Fleet List – J. & R. Griffiths

BRITANNIA (no image available)
In fleet as a sand dredger 1922-51
Built in 1906 by Fullerton, Paisley, as a steam barge.
Gross tonnage 118 – Length 90.6 feet – Beam 22.8 feet
Machinery: Steam two cylinder compound, by Hudson & Co.
History:
1906 Completed as a general cargo steam barge
1922 Purchased by J. & R. Griffiths of Cardiff; converted for sand dredging
1946 Declassed

1951 Laid up in Cardiff
1951 Sold for scrap in April; broken up by Cashmore at Newport

SKARV (no image available)
In use as a sand dredger 1940
Built in 1911 by Ardrossan DD & SB Co. as the steam coaster *Skarv*
Gross tonnage 158 – Length 90.2 feet – Beam 21.1 feet
Machinery: Steam two cylinder compound, by Ardrossan SB & DD Co.
History:
1911 Completed for Glasgow Steam Coasters Co.
1921 Owned by Suffolk Shipping of Glasgow
1939 Owned by Tyne Dock Engineering; converted to carry steel
1940 Owned or managed by J. & R. Griffiths; converted for sand dredging
1940 On a charter to the Admiralty; sails from Swansea to load sand at the Nash Sands on 11 November and disappears; Joint Arbitration Committee considers the vessel a war loss, likely sunk by a mine with the loss of her five
1951 Laid up in Cardiff
1951 Sold for scrap in April; broken up by Cashmore at Newport

INDIUM
In fleet as a sand dredger 1947-61
Built in 1932 By Charles Reynoldson & Co., South Shields, as the steam coaster *Indium*
Gross tonnage 207 – Length 110 feet – Beam 22.1 feet
Machinery: Steam two cylinder compound, by Reynoldson & Co.
History:
1923 Completed for United Alkali Co. of Liverpool
1932 Owners become Imperial Chemical Industries
1939 Sold to Shannon Steamship Co. of Kilrush
1947 Sold to J. & R. Griffiths; converted into a sand dredger; no change of name; still registered in Liverpool
1961 Sold to William France Fenwick of London
1963 Broken up in Gateshead

Newport Sand & Gravel Co.

The company was founded in Newport in April 1923 with the purchase of the ship *Isabel*, which was registered as owned by the Isabel Steamship Co. It became a tradition for Welsh ship owners to register their ships as run by a single ship company within a group of managed companies. This was to limit their liability to that ship only. If the ship owed money, that ship alone was affected and no other ships owned by the group. Alfred Tucker of Newport and Richard Griffiths of Cardiff had purchased the *Isabel* with the intention of starting an aggregate business at Newport; she was converted to a suction dredger in June 1923. J. & R. Griffiths became her management company.

She was sold for scrap in 1949, replaced by the 1915-built steam coaster *Collin*, which was renamed *Isabel*. Her owners became the Isabel Steamship Co. of Cardiff, with J. & R. Griffiths of Cardiff once more as managers. The *Isabel* mainly ran to Newport, but was also seen in Cardiff. Later, the company became a wholly owned part of J. & R. Griffiths, who in 1963 sold their aggregate dredging interests to the newly formed Western Dredgers of Newport.

Fleet List – Newport Sand & Gravel

ISABEL
In fleet as a sand dredger 1923-49
Built in 1898 by Scott of Bowling as the steam coaster *Isabel*
Gross tonnage 233 – Length 100 feet – Beam 23.1 feet
Machinery: Single screw steam compound two cylinder
History:
1898 Completed for Joseph Monks of Liverpool
1923 Sold to Alfred Tucker and Richard Griffiths
1923 Converted into sand suction dredger
1949 Sold to British Iron & Steel Corporation for scrap on 14 October

Isabel (1949-64) at Swansea.

ISABEL
In fleet as a sand dredger 1949-64
Built in 1915 by Jeffery of Alloa as the steam coaster *Collin*
Gross tonnage 287 – Length 120.5 feet – Beam 22.1 feet
Machinery: Single screw steam compound two cylinder
History:
1915 Completed for Howdens of Larne
1949 Sold to the Isabel Steamship Co.
1949 Renamed *Isabel*; converted into a sand suction dredger
1964 Sold to Tay Sand & Gravel Co. of Dundee

Swansea Sand & Gravel Co.

In 1950, William Adams set up this company in Swansea, assisted by J. & R. Griffiths of Cardiff, who initially assisted in the purchase of the steam coaster *Lynn Trader*. The *Lynn Trader* was converted for sand dredging and was renamed *Imogen* and based in Swansea. She did however also trade to Cardiff, and on rare occasions to Newport. J. & R. Griffiths's own ship, the *Indium*, based in Cardiff, also traded to Swansea. The company was first managed by J. & R. Griffiths, but later became a wholly owned subsidiary, and trading ceased when the company, together with J. & R. Griffiths, was sold to Western Dredgers of Newport in 1963. The *Imogen* was laid up at Swansea and later sold for breaking up.

Fleet List – Swansea Sand & Gravel Co.

IMOGEN (no image available)
In fleet as a sand dredger 1950-64
Built in 1927 by Yarwood & Sons as the steam coaster *Vivonia*
Gross tonnage 226 – Length 130.8 feet – Beam 23.6 feet
Machinery: Steam triple expansion, by W. J. Yarwood & Sons
History:
1927 Completed
1936 Owned by General Steam Navigation; renamed *Goldfinch*
1949 Owned by Great Yarmouth Shipping; renamed *Lynn Trader*
1950 Sold to Imogen Steamship Co. of Swansea; converted into a sand dredger and renamed *Imogen*
1963 Laid up in Swansea; offered for sale
1964 Broken up at Passage West

Sandridge & Co.

In July 1921, John George Hamlin, a tug master in the port of Cardiff, with the help of others on a share basis, purchased the steam coaster *Kyles* from owners in South Wales. In November 1921, George Hamlin had the *Kyles* converted in Cardiff into a sand suction dredger. He operated under the title of Sandridge & Co. with Hugh Hamlin, his son, as manager. Later, George Hamlin became the sole owner of the *Kyles,* although it remained under the management of Hugh Hamlin. George Hamlin worked the *Kyles* from a wharf on the bank of the Glamorganshire Canal at Harrowby Street. At this time, other sand firms also worked from Harrowby Street. These were Arthur Sessions & Son, F. Bowles & Sons, and J. & R. Griffiths.

Hamlin's main trade with the *Kyles* was to Cardiff, where he dredged his cargos of sand and gravel from East and West One Fathom Banks, near the Holms in the Bristol Channel. The *Kyles* also ran occasional cargos to Bristol; this was under an agreement with British Dredging. In 1942, Hamlin sold the *Kyles* to William Metcalf of Ilfracombe, North Devon, where she was used as a salvage vessel.

As a replacement vessel, Hamlin purchased the slightly larger and newer *Catherine Ethel* from owners in London. He immediately had her converted in Cardiff into a sand suction dredger. The company continued to run out of the Glamorganshire Canal until late on the night of 5 December 1951, when the *Catherine Ethel* collided with the lock gates at the sea lock. The lock gates were destroyed, causing the canal to drain a section over a mile long. The gates were never repaired, ending 158 years of canal operation, which allowed the City Corporation to close it to commercial traffic.

The sand dredging firms then moved to wharves in the East and West Bute Docks. F. Bowles & Sons set up a wharf on the Cardiff mud in an area known as 'the Ferry'. After his father retired, Hugh Hamlin continued to run Sandridge as a one-man, one-ship operation, and when he retired in the early 1960s, he was the last independent operator in the Cardiff sand trade. The early 1960s was a time of change for the dredging concerns using the Bute Docks; they were notified by the docks authority that the docks were to close. F. Bowles & Sons and Bristol Sand & Gravel merged to form British Dredging, and were soon joined by Arthur Sessions & Son, leaving only the berth of J. & R. Griffiths for Hamlin. With his type of operation, the tonnage was too small for the larger company to handle. When J. & R. Griffiths chose to cease trading, Hugh Hamlin retired, sold his interests to British Dredging, and sold the *Catherine Ethel* to Llanelli Quarries for further trading. When sold, the *Catherine Ethel* was among the very last coal-fired steam sand dredgers operating out of Cardiff. Time had defeated Hamlin's kind of operation.

Fleet List – Sandridge & Co.

KYLES

Kyles (1921-42) in Cumberland Basin, Bristol.

In fleet as a sand dredger 1921-42
Built in 1872 by J Fullerton & Co., Paisley, as the *Kyles*, a steam tender to the sailing fishing fleet
Gross tonnage 79.86 – Length 82.3 feet – Beam 18.2 feet
Machinery: Steam compound two cylinder, by King & Co. of Glasgow
History:
1872 Launches in March; completed as a tender to the sailing fishing fleet; used to transport fish from the fleet to the quayside markets; registered in Glasgow.
1881 Sold to William Veitch of Crieff
1886 Sold to William Taylor of Newcastle
1886 Sold to Jane Taylor and William Veitch
1889 Sold to George Saunders of Newcastle
1889 Sold to George Kent of Hull; reregistered in Hull
1900 Sold to Joseph Smith of Hull
1900 Sold to George Hiles of Hull in May
1901 Sold to John Leyshon of Pontypridd
1904 Sold to Richard Cormack Reaveney of Cardiff
1906 Re-boilered with unit built by Abbott & Co. of Newark
1911 Sold to R. Burton of Newport
1916 Sold to R. Binding of Cardiff
1919 Sold to Henry Connor of Herne Bay
1919 Sold to James Tester of Greenhithe

1920 Sold to Horace Cole and three other shareholders
1921 Sold to Alexander Gray and Thomas Smith of Treforest
1921 Sold to George Hamlin of Cardiff on 25 July
1921 Converted into sand suction dredger in November
1942 Sold to William Metcalf of Ilfracombe on 25 November; used as salvage vessel

CATHERINE ETHEL

Catherine Ethel (1942-62) at Cardiff.

In fleet as a sand dredger 1942-62
Built in 1906 by Crabtree & Co., Great Yarmouth, as the steam coaster *Mistley*
Gross tonnage 153 – Length 103 feet – Beam 19.5 feet
Machinery: Steam compound two cylinder by Crabtree & Co.
History:
1906 Completed for F. W. Horlock, barge owners in Mistley, Essex
1911 Lengthened to increase her tonnage
1920 Sold to J. Leete & Son of London; reregistered in London
1926 Sold to William H. Muller of London; works their London-Paris service
1926 Reregistered by Muller; now owned by Vianda Steam Ship Co. of London
1942 Sold to Sandridge & Co.
1942 Converted into sand suction dredger in Cardiff; cargo space cut down to prevent overloading
1962 Sold to Llanelli Quarries, with T. I. Jones as the manager

William J. Holway

William J. Holway of Cardiff was one of the single-ship operators working in the short sea coastal trade who later decided to venture into sand dredging. Not only was he a single-ship operator, he also ran the business on his own. His office was originally at Atlantic Buildings in Bute Docks, Cardiff. He later moved to Cory Buildings in Mount Stuart Square.

The company started in 1912 with the purchase of Holway's first and only ship, from owners in Greenock, Scotland. This vessel was the *Eisa*, built at Greenock in 1901. She was renamed *Lyndale* by William Holway and reregistered in Cardiff. *Lyndale* was initially used in the short sea coastal trade, but she was used in the sand trade from 1931, and converted for suction dredging in 1932.

William Holway continued as a one-man, one-ship operation until 1935, when he sold the *Lyndale* and his sand dredging interests to Arthur Sessions & Son of Cardiff.

Fleet List – William J. Holway

LYNDALE (no image available)
In fleet as a sand dredger 1931-35
Built in 1901 by G. Brown & Co., Greenock, as the steam coaster *Eisa*
Gross tonnage 138 – Length 84.1 feet – Beam 20 feet
Machinery: Steam compound two cylinder 30 rhp, by Muir & Houston, Glasgow
History:
1901 Completed for D. McEwan of Greenock; registered in Greenock
1912 Sold to William J. Holway; renamed *Lyndale*; reregistered in Cardiff; used in the short sea coastal trade
1931 Used in the sand trade
1932 Converted into a sand suction dredger
1935 Sold to Arthur Sessions & Son

George J. Binding

George Binding was one of the smaller sand dredger operators at Cardiff, venturing into ship owning in 1930 with the purchase of the small steam coaster *Gael*, from P. W. Gibb of Glasgow. Initially, this vessel was used in the short sea coastal trade out of Cardiff, and then, in 1932, she was used in the Cardiff sand trade. She remained unconverted and by 1933 was returned to the coastal trade.

It is known that by 1932 George Binding was associated with F. Bowles & Sons of Cardiff. Through them, he purchased the coaster *Jolly Esmond*, in a damaged condition, from Walford Line of London. George Binding then employed F. Bowles to repair the vessel, which he then reregistered in his own name. In 1933, F. Bowles was involved in her conversion into a sand suction dredger. On completion, she was renamed *Evelyn B*. There was a further association between George Binding and Fred Bowles when F. Bowles, through its management company, crewed and managed the vessel.

In the main, the *Evelyn B* ran to the wharves owned by both F. Bowles and Arthur Sessions. As George Binding had no berth or wharf of his own, he moved away from sand dredging in 1937, selling *Evelyn B* to Arthur Sessions & Son. The company, or rather the *Gael*, continued to run and during the period 1943-47 was registered as the property of Mrs M. Binding. There is no listing for either the *Gael* or Mrs Binding after 1947.

Fleet List – George J. Binding

GAEL
In the sand trade 1932-33
Built in 1903 by G. Brown & Co., Greenock, as the steam coaster *Gael*
Gross tonnage 108 – Length 83.4 feet – Beam 19.9 feet
Machinery: Steam compound two cylinder, by Muir & Houston, Glasgow
History:
1903 Completed to the order of P. W. Gibb of Glasgow
1930 Sold to George J. Binding of Cardiff
1932 Used in the Cardiff sand trade
1933 Returns to the short sea coastal trade
1943-47 Owner registered as Mrs M. Binding; nothing known after 1947

EVELYN B
In the fleet as a sand dredger 1933-37
Built in 1918 by N. V. V. D. Kuy & V. D. Ree, Rotterdam, for Dutch ship owners, as the steam coaster *Mies*
Gross tonnage 237 – Length 107.7 feet – Beam 21.8 feet
Machinery: Steam compound two cylinder, by the shipbuilders
History:
1918 Completed for Dutch ship owners
1920 Sold to Leopald Walford of London; renamed *Jolly Esmond*
1932 Damaged in a collision in the Thames
1932 Sold in damaged condition to George J. Binding of Cardiff; repaired; reregistered at Cardiff
1933 Converted into a sand suction dredger; renamed *Evelyn B*
1937 Sold to Arthur Sessions & Son, Cardiff; renamed *Sandale*

Western Dredgers

This company was formed in Newport in 1960. It immediately ordered new purpose-built sand dredgers from the Dutch shipyards. The first vessel was the *Isca*, delivered in 1960, followed by the *Instow* in 1964 and the *Moderator* in 1965. In 1960, the new company took over the interests of J. & R. Griffiths of Cardiff, which included interests in the sand trade of Newport Sand & Gravel and Swansea Sand & Gravel. They did not, however, take over the three ageing sand dredgers owned and managed by J. & R. Griffiths. Western Dredgers had by 1967 formed a trading agreement with British Dredging. They later found it hard to compete and were fully absorbed into the British Dredging fleet by 1976, when they were reregistered as owned by British Dredging Co. Ltd.

Fleet List – Western Dredgers

ISCA
In fleet as a sand dredger 1960-76
Built in 1960 by N. V. Scheeps, Westerbroek, Holland, as the purpose-built motor sand dredger *Isca*
Gross tonnage 550 – Length 51.67 metres – Beam 9.15 metres
Machinery: Single-screw eight-cylinder 660 bhp diesel engine, by Lister Blackstone
History:
1960 Completed for Western Dredgers, Newport
1976 Transfers to British Dredging

INSTOW
In fleet as a sand dredger 1964-76
Built in 1964 by N. V. Scheeps, Foxhol, Holland, as the purpose-built motor sand dredger *Instow*
Gross tonnage 735 – Length 186.4 feet – Beam 34.1 feet
Machinery: Single-screw eight-cylinder 800 bhp diesel engine, by Lister Blackstone
History:
1964 Completed for Western Dredgers, Newport
1976 Transfers to British Dredging

MODERATOR
In fleet as a sand dredger 1965-76
Built in 1965 by Boeles, Schps & Mchf, as the purpose-built motor sand dredger *Moderator*

Isca (1960-76) at Cardiff with the *Bowcrest*.

Instow (1964-76).

Moderator (1965-76).

Gross tonnage 836 – Length 59.44 metres – Beam 11.38 metres
Machinery: Single-screw eight-cylinder 800 bhp diesel engine, by Lister
Blackstone
History:
1965 Completed for Western Dredgers, Newport
1976 Transfers to British Dredging

British Dredging Co.

In 1962, the sand dredging business of F. Bowles & Sons merged with Bristol Sand & Gravel. This newly formed company took the title British Dredging Co. and was floated as a public company on the stock exchange. The title dates back to 1932, when it was set up to act as a non-profit management company for a pool of sand dredgers operated by five dredger owners in the Bristol Channel.

F. Bowles & Sons, prior to the merger, placed an order with Ailsa Shipbuilding of Troon for a larger purpose-built coastal aggregate dredger. This vessel, the *Bowqueen* (1,317 gross tons) was delivered in 1963, and became the first new ship for the newly formed British Dredging Co. Ltd.

During the 1960s, the company expanded its operations and spread out of the Bristol Channel to the Thames Estuary, the North East, Holland, Belgium, and France. At the same time, a massive shipbuilding programme was undertaken, together with joint venture agreements and further mergers and takeovers.

OVER THERE THE PARTHENON; and over here ...The CAMERTON!

Should you be fortunate enough to take a holiday in Greece in future years, you may be surprised to bump into an old friend. The S.S. Camerton left British Dredging Service in early November, and is now working as a sand dredger in the waters around Greece.

She was the last steam dredger in The British Dredging fleet, and was built in 1950 at Troon. An 800 ton cargo vessel, she was affectionally known as "Smokey Joe", in the Bristol Channel.

We are not sure whether there is any truth in it, but the story has it that flights into Rhoose, took bearings from the Camerton as they came over the Channel. What is true is that the Camerton has done excellent service during her time with British Dredging, and we wish her good luck with her new owners.

The Sandpiper, the house magazine of British Dredging, Christmas 1973.

Among the companies taken over were Arthur Sessions & Son of Cardiff, Western Dredgers of Newport, the Commercial Dry Dock Co. of Cardiff, and a large stakeholding in Ailsa Shipbuilding of Troon, where all the new ships would be built. British Dredging became involved in ready-mix concrete, engineering, building, and the manufacture of concrete products, as F. Bowles had done previously. At the same time, the company formed its own shipping, transport, marketing, and property development divisions. The whole of the company operations and planning were controlled from the head office in Cardiff, and in the early 1970s British Dredging was the largest company of its type in the whole of Europe. It was then operating a fleet of thirteen sand dredgers in full-time employment in the Bristol Channel, the North Sea, and on both sides of the English Channel.

The recession of the mid-to-late 1970s forced the company to dispose of a large part of its activities, and place on the disposal list some of its older vessels. A new streamlined British Dredging formed British Dredging Aggregates Ltd in 1978, to serve Cardiff and South Wales. In Bristol, a joint venture was formed with The Holms Sand & Gravel to serve Bristol and Bridgwater.

In December 1983, the company sold off its Thames, east coast and European interests to South Coast Shipping, part of the Ready Mix Concrete Group, after which these divisions were known as East Coast Aggregates Ltd.

This left British Dredging in the Bristol Channel, where it all started. In 1990, the company formed a joint venture with Ready Mix Concrete to serve the Bristol Channel, and in 2000 the company became known as RMC Bristol Channel.

Fleet List – British Dredging

BOWSTAR

Bowstar (1962-73).

In fleet as a sand dredger 1962-73
Build details: See F. Bowles & Sons Ltd
History:
1962 Transferred from the fleet of F. Bowles & Sons to British Dredging; hired out for a cable-laying contract
1972 Suffers mechanical breakdown; uneconomic to repair
1973 Towed to Dublin to be broken up

BOWLINE

Bowline (1970-75), inbound at Cumberland Bain, Hotwells, Bristol.

In fleet as a sand dredger 1970-75
Build details: See F. Bowles & Sons Ltd
History:
1962 Transfers from the fleet of F. Bowles & Sons to British Dredging
1970 Transfers within from the Cardiff fleet to the Bristol fleet, to replace the *Camerton*; trades to the Bristol and the South Wales ports
1975 Withdrawn from service and laid up at Cardiff; offered for sale
1975 Scrapped at the west end of Queen's Dock, Cardiff, late in the year

BOWCREST

Bowcrest (1962-75).

In fleet as a sand dredger 1962-75
Build details: See F. Bowles & Sons Ltd
History:
1962 Transfers from the fleet of F. Bowles & Sons to British Dredging
1963 Spends several weeks prospect-dredging in the Thames and on the North East coast
1975 Sold to Sondora Shipping, Panama; renamed *Edip*

BOWPRIDE

Bowpride (1962).

In fleet as a sand dredger 1962-62
Build details: See F. Bowles & Sons Ltd
History:
1962 Transfers from the fleet of F. Bowles & Sons to British Dredging
1962 Transfers to British Dredging's Thames and East Coast fleet to work out
of the River Humber
1970 Springs a leak and sinks in the lower Humber on 11 January

DUNKERTON

Dunkerton (1962-65) at Newport.

In fleet as a sand dredger 1962-65
Build details: See Bristol Sand & Gravel Co. (*Volume One: The English
Coast*)
History:
1962 Transfers from the fleet of Bristol Sand & Gravel; mainly based in South
Wales, but remains in the Bristol fleet
1965 Sold for scrap; broken up by John Cashmore in Newport

CAMERTON
In fleet as a sand dredger 1962-73
Build details: See Bristol Sand & Gravel Co. (*Volume One: The English
Coast*)
History:
1962 Transfers from the fleet of Bristol Sand & Gravel

1970 Transfers within British Dredging from the Bristol fleet to the Cardiff fleet; replaced by the *Bowline*
1973 Sold to Roussos Brothers, Greece; renamed *Archonto*

BADMINTON

Badminton (1962-75) in the River Avon by Bridge Valley Road, Bristol.

In fleet as a sand dredger 1962-75
Build details: See Bristol Sand & Gravel Co. (*Volume One: The English Coast*)
History:
1962 Transfers from the fleet of Bristol Sand & Gravel; works mainly out of the South Wales ports and Bridgwater, with occasional visits to Bristol
1963 Grounded on pack ice near Dunball, Bridgwater, on 9 February, badly damaging her bottom plating; inspected in Bristol; sails to Ailsa of Troon (part-owned by British Dredging) for extensive repairs; gains an almost completely new bottom
1975 Sold to Louis Piriou, St Malo; renamed *Le Roselier*
1980s Broken up

PETERSTON

Peterston (1962-65, 1976-78) at Barry.

In fleet as a sand dredger 1962-65, 1976-78
Build details: See Bristol Sand & Gravel Co. (*Volume One: The English Coast*)
History:
1962 Transfers from the fleet of Bristol Sand & Gravel
1965 Transfers to British Dredging's London fleet
1976 Returns to British Dredging's Bristol fleet
1978 Transfers to British Dredging Aggregates

ISCA
In fleet as a sand dredger 1967-77
Build details: See Western Dredgers Ltd, Newport
History:
1967 Transfers from the fleet of Western Dredgers of Newport to British
Dredging; works in the Bristol Channel; mainly serves ports in South Wales
1976 Western Dredgers of Newport sold to British Dredging
1977 Sold to Louis M. Piriou, Paimpol; renamed *Le Ferlas*

INSTOW

Instow (1967) at Dover.

In fleet as a sand dredger 1967
Build details: See Western Dredgers Ltd, Newport
History:
1967 Transfers from the fleet of Western Dredgers of Newport to British Dredging
1967 Transfer to British Dredging's Thames and East Coast fleet

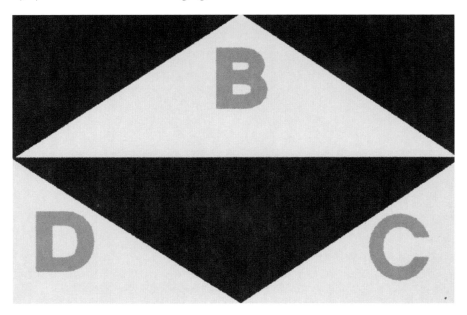

British Dredging Company Ltd logo.

MODERATOR
In fleet as a sand dredger 1967
Build details: See Western Dredgers Ltd, Newport
History:
1967 Transfers from Western Dredgers of Newport to British Dredging
1967 Transfers to British Dredging's Thames and East Coast fleet

BOWQUEEN

Bowqueen (1974-78) at Cardiff.

In fleet as a sand dredger 1974-78
Built in 1963 by Ailsa Shipbuilding, Troon, as the purpose-built motor sand
dredger *Bowqueen*
Gross tonnage 1,317 – Length 257.7 feet – Beam 39.1 feet
Machinery: Eight-cylinder 1,864 bhp diesel engine, by Mirrlees National
History:
1963 Becomes the first ship built for the newly formed British Dredging
1963 Based in the British Dredging Thames and North East fleet
1965 Capsizes in a severe gale off Walton-on-the-Naze on 8 September, with
the loss of four crewmembers

1965 Raised and towed to Holland for repair on 10 October
1974 Transfers to the British Dredging's Cardiff fleet
1978 Transfers to British Dredging Aggregates

BOWCROSS

Bowcross (1970-78) at Cardiff.

In fleet as a sand dredger 1970-78
Built in 1967 by Goole Ship Building as the purpose-built motor sand dredger
Chichester Cross
Gross tonnage 959 –Length 59.8 metres – Beam 12 metres
Machinery: Eight-cylinder 1,000 bhp diesel engine by Mirrlees National
History:
1967 Completed for John Heaver of Chichester
1970 Purchased by British Dredging; renamed *Bowcross*
1971 Commences work in the Bristol Channel; trades to Bristol, Bridgwater, and all the South Wales ports
1978 Trades only to South Wales ports from February
1978 Transfers to British Dredging Aggregates in November

Burry Sand Company

Burry Sand Co. Ltd was formed at Llanelli in 1965 by the local building contractor Isaac Jones, who operated his building business from a yard in an area known locally as the Furnace. Burry Sand was one of the smaller aggregate dredging firms operating in the Bristol Channel. It did, however, operate some interesting ships, and through various changes of ownership the company finally became part of the RMC Group. Isaac Jones established his sand-dredging business working out of the old Carmarthen Dock in Llanelli. Cargos of sand were dredged from the estuary outside Llanelli and from the near sand banks in the Bristol Channel. Later, the larger dredgers worked at the Helwick Sands.

Isaac Jones purchased a vessel locally in 1964 and converted her for sand dredging in 1965; she was the motor barge *Carry*, which had been built in Bristol in 1933 to the order of the Severn and Canal Carrying Co. In 1965, he purchased a newly converted, lengthened, re-engined sand dredger from Appledore Shipbuilders, which he renamed *Coedmor*. This vessel had started her life in 1946 as the steam-powered *VIC 57*, an Admiralty victualling inshore craft that bore little resemblance to what she would become. In 1968, Burry Sand again expanded, with the purchase and conversion of the motor coaster *Orselina*. With the arrival of this vessel, the *Carry* was laid up, and in the early 1970s she was broken up at Llanelli by Isaac Jones at the old Richard Thomas and Baldwins Wharf outside the Carmarthen Dock. The *Orselina* worked for the company on a regular basis until she was broken up in 1972 by Isaac Jones. Burry Sand continued to trade using the *Coedmor* until 1978, when the company was sold to Llanelli Plant Hire. Isaac Jones building contractors later went into liquidation, and were taken over by the Mowlem Group.

Llanelli Plant Hire Co.

In 1978, the interests of Burry Sand were taken over by Llanelli Plant Hire Co. Ltd and the name Burry Sand was dropped. The new company replaced the *Coedmor* with a newly purchased ship, the *Rhone*, which they had converted into a sand dredger. This vessel, built in Holland in 1966, had a gross tonnage of 276. The increased tonnage allowed the company the flexibility of operating one ship. The company continued to trade from Llanelli, but also started to work out of Swansea, where the *Rhone* had her port of registry. In 1982, the company was sold to new owners, who moved the operation to Swansea and took the title Bury Sand Co. Ltd. Note that 'Bury' had now dropped an 'r'.

Bury Sand Co.

Under its new title and ownership, Bury Sand moved its place of business and operations from Llanelli to Swansea during the early part of 1982; from then on its vessel the *Rhone* was to be seen trading in and out of Swansea and on occasion to Briton Ferry. Although still operating out of Swansea, the company had a change of address for a short time in 1985, and became Bury Sand of Bridgend, Glamorgan. It was also in 1985 that the company was offered for sale, and was bought out by Ready Mix Concrete.

In September 1985, the *Rhone* was sold to Mr R. Brean, who later operated the vessel out of Briton Ferry as Cross Avon Ltd. He also operated the vessel from Newport as Severn Sands Ltd.

The replacement vessel for Bury Sand came from Ready Mix Concrete's south coast fleet in Southampton. It was the 1964 purpose-built motor sand dredger *Sand Tern*, which retained her RMC funnel colour and port of registry at Southampton.

Bury Sand continued to trade as part of Ready Mix Concrete until March 1990, when Ready Mix Concrete entered into a joint agreement with British Dredging Aggregates of Cardiff to serve the Bristol Channel.

After this, both Bury Sand and the *Sand Tern* effectively became part of British Dredging Aggregates, and the *Sand Tern* was repainted with the new company's colours. In 2002, the company became known as RMC Bristol Channel Ltd.

Fleet List – Burry Sand, Llanelli Plant Hire, Bury Sand

CARRY

Carry (1964-67).

In fleet as a sand dredger 1964-67
Built in 1933 by Charles Hill & Sons, Bristol, as the general cargo motor
barge *Severn Carrier*
Gross tonnage 101 – Length 89.6 feet – Beam 19.6 feet
Machinery: Four-cylinder diesel engine, by Petter Ltd, Yeovil
History:
1933 Completed to the order of the Severn & Canal Carrying Co.
1948 Owners become Lyon & Lyon Ltd
1957 Sold to John Alt for use in the Bristol coal trade; renamed *Carry*
1962 Sold to Peter Evans of Llanelli
1964 Owner becomes Isaac Jones; converted into a sand dredger
1967 Laid up at Llanelli
1970s Broken up at Llanelli; engine salvaged and returned to Bristol

COEDMOR

Coedmor (1965-78) alongside *Orselina* (1968-72) at Llanelli.

In fleet as a sand dredger 1965-78
Built in 1946 by J. Pollock & Sons, Faversham, as the steam-powered
victualling inshore craft *VIC 57*
Gross tonnage 147 – Length 80 feet – Beam 20 feet
Machinery: Compound two cylinder steam, by Pollock
History:
1946 Completed for the Ministry of War Transport
1948 Sold to Arran Sea Transport, Bute; renamed *Arran Monarch*
1953 Sold to Wansbrough Paper Co. of Watchet, Somerset

1960 Sold to Peter Herbert of Bude
1963 Sold to Appledore Shipbuilders; lengthened and converted into a diesel sand dredger
1965 Sold to Burry Sand Co; renamed *Coedmor*
1978 Sold back to Peter Herbert

ORSELINA

Orselina (1968-72) at Swansea.

In fleet as a sand dredger 1968-72
Built in 1938 by A. Vuijk & Zonen, Capelle, as the motor coaster *Brixham*
Gross tonnage 258 – Length 136.6 feet – Beam 23.7 feet
Machinery: Five-cylinder 320 bhp diesel engine, by Crossley Bros
History:
1938 Completed for H. Harrison Shipping of London
1940 Sold to R. Rix & Son; renamed *Ebrix*
1960 Sold to Hazely of Guernsey; renamed *Orselina*
1966 Sold to Continental Cargos of London
1968 Sold to Burry Sand; converted into a sand dredger
1972 Broken up in Llanelli by Isaac Jones

TONY (no image available)
Built in 1925 by Richard Dunston, Thorne, as a motor tank barge for John Harker Ltd
Gross tonnage 55 – Length 73.5 feet – Beam 14 feet
History: Sold to owners in Llanelli; converted for sand dredging; nothing further is known

TANKARD X8

Tankard X8 (1961-68) at Llanelli.

In fleet as a sand dredger 1961-68
Built in 1915 by Sir James Laing & Sons, Sunderland, as an X-class lighter for
the Admiralty
Gross tonnage 134 – Length 110 feet – Beam 21 feet
Machinery: Four-cylinder 160 bhp oil engine, by Blackstone
History:
1915 Completed; based in Dover and Sheerness
1922 Sold to Vauxhall Trading, Sheerness
1927 Owned by BP Tanker Co.; used as a freshwater barge
1950 Sold to Prince of Wales Dry Dock Swansea; metal trading in Swansea
1961 Sold to George Tate of Swansea; becomes a sand dredger in Llanelli
1974 Still working; nothing further known

RHONE

Rhone (1978-85) at Swansea.

In fleet as a sand dredger 1978-85
Built in 1966 by Schps, Appingedam, as the motor coaster and explosives
carrier *Rhone*
Gross tonnage 276 – Length 46.36 metres – Beam 7.62 metres
Machinery: Six-cylinder 400 bhp diesel engine, by Kromhout Motorenfab
History:
1966 Completed for Rederij Sambre NV Kamps of Groningen
1978 Sold to Llanelli Plant Hire; converted into a sand dredger
1982 Transfers as a part of Bury Sand
1985 Sold to Mr R. Brean of Newport

SAND TERN

Sand Tern (1985-90).

In fleet as a sand dredger 1985-90
Built in 1964 by P. K. Harris of Appledore as the purpose-built motor sand
dredger *Sand Tern*; finished by J. Bolson of Poole
Gross tonnage 540 – Length 174.1 feet – Beam 30.4 feet
Machinery: Six-cylinder diesel engine, by Lister Blackstone Marine
History:
1964 Completed to the order of South Coast Shipping; services the south
coast of England
1972 Owners become William Cory and Ready Mix Concrete
1985 Transfers to Bury Sand
1990 Transfers to British Dredging Aggregates

British Dredging Aggregates

In December 1978, the Cardiff-based side of British Dredging plc became
British Dredging Aggregates Ltd.

In March 1990, British Dredging and Ready Mix Concrete entered into
a joint-ownership agreement to operate in the Bristol Channel as British
Dredging Aggregates Limited.

The *Sand Tern* from Bury Sand in Swansea was repainted in British Dredging
Aggregates colours in the spring of 1990.

On 5 November 1992, British Dredging Aggregates re-established their link
with the sand trade at Avonmouth with the opening of a new depot at M
Berth in the old dock.

As this berth at Avonmouth had no cranes, it would be supplied by British Dredging Aggregates ship *Welsh Piper*, which was capable of arriving on the first of the tide, discharging herself and sailing on the last of the same tide.

In 2000, the company became known as RMC Bristol Channel. In 2002, the whole group became RMC Marine Ltd.

The group was bought out by Cemex SA of Mexico in 2005 and now operates as Cemex UK Marine Ltd.

Only two ships regularly call at the wharf, both of which are capable of self-discharge. For many years, the only regular visitor was the *Welsh Piper*, which single-handedly kept the wharf supplied. Cemex then had the *Sand Serin* converted to self-discharge. She appeared from 2007, when the *Welsh Piper* was either working away or under repair.

This advertisement, from 5 Nov. 1992, announces the opening of M Berth in Avonmouth.

Fleet List – British Dredging Aggregates, Ready Mix Concrete, Cemex

PETERSTON

Peterston (1978-89) at Cardiff.

In fleet as a sand dredger 1978-89
History:
1978 Transfers to British Dredging Ltd
1989 Becomes reserve ship; windlass and generator removed and fitted to
Bowcross to keep her serviceable
1991 Towed to Newport; broken up

BOWCROSS

Bowcross (1978-99) at Pooles Wharf, passed by a fully loaded *Sand Pearl*.

In fleet as a sand dredger 1978-99
History:
1978 Transfers to British Dredging Ltd in November
1989 Windlass and generator fitted (from the *Peterston*)
1998 Laid up in Cardiff; offered for sale
1999 Sold to foreign interests; sails from Cardiff; renamed *Rita 1*

BOWQUEEN

Bowqueen (1978-88) on the River Taff at Ferry Wharf, Cardiff.

In fleet as a sand dredger 1978-88
History:
1978 Transferred from British Dredging Ltd, Cardiff
1988 Purchased by Madeiran owners on 5 August; towed from Cardiff by the tug *Avongarth*, bound for Madeira via Lisbon; renamed *Susana Christina* at Funchal

SAND TERN
In fleet as a sand dredger 1990-98
History:
1990 Transfers from Bury Sand, Swansea, to British Dredging Aggregates in March
1990 Repainted in British Dredging Aggregate colours in the spring; little used; laid up in Cardiff and then Barry
1998 Towed from Barry by the tug *Towing Chieftan* on 19 November, bound for a Belgian shipbreaker

WELSH PIPER

Welsh Piper (1987-present) at Newport.

In fleet as a sand dredger 1987-present
Built in 1987 by Appledore Shipbuilders as the purpose-built motor sand dredger *Welsh Piper*; capable of self-discharge with bucket scrappers and conveyor
Gross tonnage 1,251 – Length 69.02 metres – Beam 12.5 metres
Machinery: Six-cylinder 1,347 bhp diesel engine, by Mirrlees Blackstone
History:
1987 Completed to the order of Filbuck 111, a British Dredging Ltd holding company
1990 Part of a joint operating venture between British Dredging Aggregates and Ready Mix Concrete
2000 Owners restyled as RMC Marine Bristol Channel
2002 Owners become RMC Marine
2005 Owners become Cemex UK Marine

SAND SERIN

Sand Serin (2002-present) at Briton Ferry.

Built in 1974 by Cleland Shipbuilding, Wallsend, as the purpose-built motor sand dredger *Sand Serin*; acts as a relief ship for the *Welsh Piper*
Gross tonnage 1,283 – Length 65.7 metres – Beam 12.4 metres
Machinery: Single-screw eight-cylinder 1,150 bhp diesel engine, by Mirrlees Blackstone
History:
1974 Completed to the order of Baring Bros Container and Leasing Co.; delivered on 20 December; built as a crane-discharge vessel; converted to self-discharge to shore via conveyor
2002 Owners become RMC Marine
2005 Owners become Cemex UK Marine

United Marine Aggregates

United Marine Agreggates logo.

The origins of United Marine Aggregates date to the formation of the Hoveringham Group in 1959. The group took its name from a village in Nottingham where they excavated gravel from land-based gravel pits. Marine dredging in the Bristol Channel by Hoveringham started in 1965, when they bought out the interests in the Sand & Gravel Marketing Co. of Cardiff. This company was itself formed in 1947, through the purchase of the Channel Sand & Ballast Co. of Swansea, to supply much-needed sand for the post-war rebuilding of Swansea and South Wales. Although based in Swansea, Channel Sand & Ballast operated throughout the South Wales ports.

On the south coast, John Heaver Ltd – formed in 1967 – joined forces with Francis Concrete of Chichester in 1981 to form Tarmac Sand & Gravel Ltd. In 1984, both the Bristol Channel and south coast operations were merged to form Tarmac Marine, which at the same time took over the interests of South Wales Sand & Gravel at Swansea, formed in 1932. In 1987, Tarmac and Pioneer Concrete entered into a joint venture agreement. The result of this agreement was the formation of United Marine Aggregates Ltd (UMA).

South Wales Sand & Gravel Co.

South Wales Sand & Gravel Co. Ltd was formed in Swansea in 1932 by members of the Bevan family. The company – through Randolf Bevan, then his brother Llewelyn, and finally Llewelyn's son John – was to remain family-owned and family-controlled for over fifty years until it was sold in 1984.

Their first ship, the 1918-built steam coaster *Glen Helen*, was purchased and converted for sand dredging in 1932. Cargos of aggregate dredged from the Bristol Channel were discharged at Swansea's North Dock, where the company set up a wharf and depot. Later, the company also operated from a wharf at Briton Ferry, which became known locally as Bevan Port.

A steady business was built up by the company in and around Swansea and, with further demand, the company purchased and converted a second steam coaster in February 1939: after conversion, this vessel took the name *Glen Spray*. All future ships owned by South Wales Sand & Gravel bore the prefix 'Glen' in their names, and later became known in the Bristol Channel as the Swansea Glens.

The company continued to operate through the Second World War, even though Llewelyn Bevan was called to serve in the army, rising to the rank of colonel by the time he returned to Swansea and rejoined the family business.

The post-war period saw South Wales Sand & Gravel working hard to supply aggregate for the rebuilding boom. In January 1948 a third steam coaster was purchased and converted for sand dredging; on completion, she took the name *Glen Foam*.

The company continued to operate their three elderly steam dredgers through the 1950s and into the 1960s. These ships, although old and steam-powered, were kept in as perfect a condition as possible.

The company ordered and took delivery of a new purpose-built motor dredger in 1960. Named *Glen Hafod*, she had been built to order by Smit & Zoon at Foxhol in Holland.

Faced with finding a new berth and wharf with the closure of the North Dock at Swansea, the company's vessels were to be seen operating from the corner of the South Dock basin and from the Prince of Wales Dock. They also continued to use their wharf at Briton Ferry.

In the early 1960s, South Wales Sand & Gravel were among the last few operators using steam ships. Difficulties in the regular supply of good-quality firing coal and skilled firemen led to the demise of the steam ships in the mid-1960s.

The company returned to Smit & Zoon for what was to be its last ship. Again, she was a purpose-built motor ship and a near sister to the *Glen Hafod*. This ship, the *Glen Gower*, was delivered from her builders in 1963. With two modern motor ships in service, the steam ships were sold – the new ships, with fifty per cent extra tonnage, had soon shown their worth and versatility.

The arrival of the *Glen Gower* allowed South Wales Sand & Gravel to supply both Swansea and Briton Ferry with sand and gravel. Cargos were also taken up-channel as far as Chepstow, where the company had built up a small trade.

The company had a second and more interesting trade into their wharf at Briton Ferry. Here, the *Glens* also discharged cargos of coal that had been dredged in the Bristol Channel from a coal seam off Barry.

As with other aggregate dredging concerns, which were largely dependent on the building industry, the Bevan family saw both boom times and times of recession within the industry. After fifty years in business, the family decided to sell South Wales Sand & Gravel and its two remaining ships.

In 1984, the company was sold and became part of Tarmac Roadstone Holdings, together with the *Glen Hafod* and the *Glen Gower*, which had worked together for over twenty-one years. The new company sold the *Glen Gower* in late 1984 to Northwood Fareham Shipping. Initially, she ran to the Northwood Pioneer concrete plant. Later, she was to be seen trading extensively in the Solent area, still as the *Glen Gower*.

The *Glen Hafod* was sold late in 1986 to Kendall Brothers of Portsmouth, who in 1988 renamed her *KB 11*. It is interesting to note that at Easter 1988, she was seen at Portsmouth bearing both names on her stern.

Fleet List – South Wales Sand & Gravel

GLEN HELEN

Glen Helen (1932-61) at Swansea.

In fleet as a sand dredger 1932-61
Built in 1918 by Crabtree & Co., Great Yarmouth, as the steam coaster *Mary Aiston*
Gross tonnage 315 – Length 130.5 feet – Beam 23.2 feet
Machinery: Steam compound two cylinder 45 rhp, by Crabtree & Co.
History:
1918 Completed to the order of W. Aiston of Scarborough
1920-21 Sold to Wilson Bros; renamed *Glen Helen*
1924 Owners become Wilson Bros, Robbin & Co.
1932 Sold to South Wales Sand & Gravel; converted into a sand suction dredger
1961 Sold to Mrs J. McLennan of Dundee

GLEN SPRAY

Glen Spray (1939-62) at Swansea.

In fleet as a sand dredger 1939-62
Built in 1921 by Crabtree & Co., Great Yarmouth, as the steam coaster *Mary Aiston 11*
Gross tonnage 305 – Length 130 feet – Beam 23.2 feet
Machinery: Steam compound two cylinder, by Crabtree & Co.
History:
1921 Completed to the order of W. Aiston of Scarborough
1928 Sold to Wallace Brothers Mersey SS Co.; renamed *Broswall*
1934 Sold to G. Couper & Co. of Liverpool; renamed *Halladale*
1939 Sold to South Wales Sand & Gravel; converted into a sand suction dredger; renamed *Glen Spray*
1962 Sold to Davies, Middleton & Davies of Cardiff; renamed *Insitu*

GLEN FOAM
In fleet as a sand dredger 1948-64
Built in 1920 by J. & D. Morris, Newcastle, as the steam coaster *Mickleham*
Gross tonnage 366 – Length 143 feet – Beam 23.8 feet
Machinery: Steam compound two cylinder, by Shields Engineering & Dry Dock Co.
History:
1920 Completed to the order of John Harrison of London
1922 Sold to Leopald Walford Shipping London; renamed *Jolly Laura*
1928 Sold to Tyne Tees Shipping; renamed *Belford*
1937 Sold to J. M. Piggins of Montrose; registered in Montrose
1948 Sold to South Wales Sand & Gravel in January; converted into a sand suction dredger; renamed *Glen Foam*
1963 The last steam sand dredger working for South Wales Sand & Gravel
1963 Sold to T. W. Ward at Briton Ferry for scrapping; arrives 22 October after being laid up for six months in Swansea

GLEN HAFOD

Glen Hafod (1960-84) entering Swansea Docks.

In fleet as a sand dredger 1960-84
Built in 1960 by Smit & Zoon, Foxhol, as the purpose-built motor sand dredger *Glen Hafod*
Gross tonnage 552 – Length 169.7 feet – Beam 30 feet
Machinery: Single-screw eight-cylinder 660 bhp diesel engine by Lister Blackstone
History:
1960 Completed to the order of South Wales Sand & Gravel; trades mainly to Swansea and Briton Ferry, on occasions to other South Wales ports and up the Bristol Channel to Chepstow
1984 Transfers to Tarmac Roadstone Holdings

GLEN GOWER

Glen Gower (1963-84) outward from Chepstow on the River Wye.

In fleet as a sand dredger 1963-84
Built in 1960 by Smit & Zoon, Foxhol, as the purpose-built motor sand dredger *Glen Gower*
Gross tonnage 552 – Length 169.7 feet – Beam 30 feet
Machinery: Single-screw eight-cylinder 660 bhp diesel engine by Lister Blackstone
History:
1963 Completed to the order of South Wales Sand & Gravel; trades mainly to Swansea and Briton Ferry, on occasions to other South Wales ports and up the Bristol Channel to Chepstow
1984 Transfers to Tarmac Roadstone Holdings

Davies, Middleton and Davies

Little is known about this company, except that it was associated with DMD Plant Hire, who in 1962 set up a concrete company known as InSitu Concrete, which was based at Cardiff. It is thought that the company branched out into sand dredging to supply their own concrete company; the company appear to have survived in sand dredging for almost a year.

Their first dredger was the 3,145-ton ex-Mersey dredger *Hoyle*, which the company purchased in 1962 and sailed to Cardiff. On arrival, she was reregistered in Cardiff, and renamed *Sand Galore*.

The company soon found this vessel too large for their requirements, and in late 1962, she was sold to a London-based company that had formed a sand-dredging business based in the Bristol Channel; this company was Sand & Gravel Marketing.

A replacement sand dredger was purchased from South Wales Sand & Gravel in Swansea. She was the *Glen Spray*, which they took to Cardiff and renamed *Insitu* after the name of the company. She was reregistered in Cardiff.

The company traded for less than twelve months; it sold its sand dredging interests and InSitu Concrete to Sand & Gravel Marketing of London.

Channel Sand & Ballast Co.

Channel Sand & Ballast Co. was formed in Swansea towards the end of 1947, when the two partners, known locally in the Bristol Channel as Mack and Ernie, purchased an ex-Government landing craft tank (or LCT). This vessel was taken to Holland and during early 1948 she was converted into a sand suction dredger; on completion she was named *Sandmoor*. Based in Swansea, she worked the South Wales ports. At the time this company was formed, there was a shortage of suitable coastal tonnage of right the type and size for conversion into sand dredgers – hence the strange choice of vessel. When the conversion was completed, the *Sandmoor* was for a short time the largest sand dredger working on the Bristol Channel. Her base was the South Dock in Swansea.

By the time Channel Sand & Ballast arrived on the scene, there were already well established sand dredging companies working in the Bristol Channel and out of Swansea. It is thought shortages in the supply of aggregate for the post-war rebuilding of South Wales made an opening for the company – not only in Swansea, but later in Cardiff and Newport. It is of interest to note that even after the extensive alterations and reconstruction, she was still recognisable as a former landing craft. After her conversion, she remained a triple-screw vessel with a main triple-expansion steam engine, and two diesel engines driving her port and starboard propellers.

Channel Sand & Ballast was soon able to show, with the *Sandmoor*, the advantages to be gained by having a vessel able to load and deliver twice the average amount of cargo – 850 tons. During the 1950s, many of the established aggregate companies invested in new purpose-built larger tonnage, but none matched the *Sandmoor* for sheer size and power. *Sandmoor* must surely rate as the most interesting vessel ever to be converted for Bristol Channel sand dredging.

Channel Sand & Ballast continued to trade until 1962 when the partners put the business up for sale. The company was taken over by the newly formed Sand & Gravel Marketing, who were based in Cardiff with offices in London. The *Sandmoor* was laid up at Swansea in early 1963, not part of the sale of Channel Sand & Ballast. By March, her survey was overdue, and in June she was sold to John Cashmore of Newport for breaking up.

Sand & Gravel Marketing

Sand & Gravel Marketing was formed in Cardiff in the latter part of 1962, when an interest was obtained in the Channel Sand & Ballast Co. of Swansea. Working out of Cardiff, the company served the South Wales ports of Newport, Cardiff, and Swansea. Although its operations were in South Wales and it was based in Cardiff, the company's registered head office was 6 St James's Square, London.

The company purchased and used two elderly secondhand dredgers in the Bristol Channel, which was to remain their only base of operation; these dredgers were the *Sand Galore* and the *Insitu*, which were part of the purchase of Davies, Middleton & Davies of Cardiff. It is understood, and there seems little doubt, that Channel Sand & Ballast of Swansea took an active interest in setting up Sand & Gravel Marketing, and by the time the *Sand Galore* had commenced work, Channel Sand & Ballast ship *Sandmoor* was laid up and part of her crew had joined the *Sand Galore*.

However, Channel Sand & Ballast continued to trade for a short time using the *Bowline*, which they hired from British Dredging. They also ran cargos for them to Newport and Cardiff using the *Dunkerton*; during this time, the *Bowline* was mainly based in Swansea. Some time during the early part of 1963, the interests of Channel Sand & Ballast were bought out by Sand & Gravel Marketing, who now found themselves with wharves at Newport, Cardiff, and Swansea that required regular supplies of aggregates.

The new company found that the *Sand Galore* was too large, unreliable, slow and and generally unsuitable for the Bristol Channel sand trade. The company began hiring ships from rival fleets in order to keep pace with demand. Sand & Gravel Marketing was unlikely to become a rival to the well-established major operators in the Bristol Channel, but the company did demonstrate how a company could operate in the sand trade with two very different ships.

In the early part of 1965, Sand & Gravel Marketing sold its Bristol Channel sand dredging interests to a company based in Nottingham who were involved in the dredging of gravel from land-based pits; this new company was known in the Bristol Channel as Hoveringham Gravels Ltd.

Fleet List – Davies, Middleton and Davies, Channel Sand & Ballast, Sand & Gravel Marketing

SAND GALORE
In fleet as a sand dredger 1962-65
Built in 1935 by Cammell Laird & Co., Birkenhead, as the mud dredger/hopper *Hoyle*
Gross tonnage 3,145 – Length 341.11 feet – Beam 54.4 feet
Machinery: Steam triple expansion six cylinder, by Cammell Laird
History:
1935 Completed for the Mersey Docks and Harbour Board
1962 Sold to Davies, Middleton & Davies of Cardiff; renamed *Sand Galore*; reregistered in Cardiff
1962 Sold to Sand & Gravel Marketing, London
1965 Sold to Hoveringham Gravels

INSITU

Insitu (1962-64) at Swansea.

In fleet as a sand dredger 1962-64

Build details: See South Wales Sand & Gravel Co. Ltd

History:

1939 Converted to sand suction dredger and renamed *Glen Spray*

1962 Sold to Davies Middleton & Davies, Cardiff. Renamed *Insitu* reregistered at Cardiff

1963 Sold to Sand & Gravel Marketing Co., London, for service in the Bristol Channel; retains the name *Insitu*

1964 Delivered to ship breakers at Passage West, Cork for scrapping in April

SAND MOOR

Sand Moor (1947-63).

In fleet as a sand dredger 1947-63

Built in 1944 by Stockton Engineering & Construction, Stockton, as the landing craft tank 'LCT (2) No. 454'

Gross tonnage 851 – Length 229.4 feet – Beam 38.2 feet

Machinery: Triple screw triple expansion steam on centre shaft, Twin Whites diesels on the port and starboard shafts, by Stockton Engineering & Construction

History:

1944 Completed for the British Government as LCT (2) No. 454

1947 Sold to Channel Sand & Ballast, Swansea

1948 Converted into a sand suction dredger in Holland; works mainly out of Swansea, also Cardiff and Newport

1963 Laid up in Swansea; surveys overdue by March
1963 Sold to John Cashmore of Newport for scrapping in June

Hoveringham Gravels

Hoveringham Gravels took its name from the gravel pits at Hoveringham, near Nottingham. Here, the Hoveringham Group dredged for, and extracted, pit gravel from lakes and open-cast, shore-based sites.

The Hoveringham Group was well established around its home base, and in 1959/60, it moved into marine aggregate dredging in the Mersey area; initially taking an interest in, and later buying out, Richard Abels & Son of Liverpool.

The group's dredging operations in the Bristol Channel date back to 1965, when they bought out the interests of the Sand & Gravel Marketing, and in so doing made it a further part of their marine aggregate dredging division. In addition to a wharf at Cardiff, Sand & Gravel Marketing also had wharves at Newport and Swansea; they also used a common-user wharf at Briton Ferry. All these wharves had in the main been supplied by Sand & Gravel Marketing's elderly sand dredger *Sand Galore*, an ex-Mersey mud dredger; additional cargos were supplied as required by other operators.

The new company formed in this buy-out took the name Hoveringham Gravels Ltd, and in 1965 it was registered as part of the Hoveringham Group. The new company found the *Sand Galore* too large to operate successfully. Due to her age, she was becoming obsolete.

After due and careful consideration, an order was placed with Appledore Shipbuilders for a purpose-built motor sand dredger capable of self-discharge. This new vessel, delivered from her builders in 1966, took the name *Hoveringham 1*. When delivered, she was to be the first vessel in the UK to employ scraper buckets in her self-discharge system. These buckets, operated by wire ropes, were dragged down a ramp into the sand tank, where sand was scooped into the bucket. This was then hauled back up the ramp and tipped onto a conveyor, which then transported the cargo to shore installations.

With the *Hoveringham I* in service, the *Sand Galore* was withdrawn and sold for scrap. It was on *Hoveringham I* that the group's colours were first seen in the Bristol Channel; both the flag and funnel were orange with a large black mammoth – a familiar sight in the Bristol Channel in the years to come.

The group further expanded to the Thames and to the north-east. It converted two motor coasters into sand dredgers; these ships took the names *Hoveringham II* and *Hoveringham III*.

Later, two larger versions of the *Hoveringham 1* were built, which after completion were returned to their builders and lengthened. These ships took the names *Hoveringham IV* and *Hoveringham V*. The last ship built for the company was of an entirely new concept, and took the name *Hoveringham VI*.

The only Hoveringham ships regularly used in the Bristol Channel were *Hoveringham I*, *Hoveringham IV*, and *Hoveringham VI*. In 1984, Hoveringham sold their marine aggregate dredging interests to Tarmac Roadstone Holdings.

Hoveringham Gravels' mammoth logo.

Fleet List – Hoveringham Gravels

SAND GALORE

Sand Galore (1965-66) near Swansea.

In fleet as a sand dredger 1965-66
1965 Purchased from Sand & Gravel Marketing
1966 Sold to T. W. Ward; scrapped in Preston

HOVERINGHAM I

Hoveringham I (1966-84) at Cardiff.

In fleet as a sand dredger 1966-84
Built in 1966 by Appledore Shipbuilders as the purpose-built motor sand dredger *Hoveringham I*
Gross tonnage 897 – Length 203.11 feet – Beam 37.1 feet
Machinery: Single-screw seven-cylinder 890 bhp diesel engine, by Ruston & Hornsby
History:
1966 Completed to the order of Hoveringham Gravels; built for service in the Bristol Channel; capable of self-discharge using bucket scrapers
1984 Ownership transfers to Tarmac Roadstone Holdings

HOVERINGHAM IV

City of Bristol, formerly *Hoveringham IV* (1969-84), at Avonmouth.

In fleet as a sand dredger 1969-84
Built in 1969 by Appledore Shipbuilders as the purpose-built motor sand
dredger *Hoveringham IV*, a sister ship to *Hoveringham V*
Gross tonnage 1,027 – Length 72.01 metres – Beam 12.07 metres.
Machinery: Single-screw eight-cylinder 1,050 bhp engine. by English Electric
Diesels
History:
1969 Completed to the order of Hoveringham Gravels
1973 Returned to her builders and lengthened
1984 Ownership transfers to Tarmac Roadstone Holdings

HOVERINGHAM V

Hoveringham V (1969-84) with *Hoveringham I* at Cardiff

In fleet as a sand dredger 1969-84
Built in 1969 by Appledore Shipbuilders as the purpose-built motor sand
dredger *Hoveringham V*, a sister ship to *Hoveringham IV*
Gross tonnage 1,027 – Length 72.01 metres – Beam 12.07 metres
Machinery: Single-screw eight-cylinder 1,050 bhp engine by English Electric
Diesels
History:
1969 Completed to the order of Hoveringham Gravels
1973 Returned to her builders and lengthened
1984 Ownership transfers to Tarmac Roadstone Holdings

Tarmac Marine

Tarmac Roadstone Holdings were well established as quarry owners, stone suppliers, and road surfacers. They also had interests in both industrial and domestic building by the time they started marine aggregate dredging in the Bristol Channel. The company first became involved in marine aggregate dredging in Chichester.

On the south coast, John Heaver Ltd, formed in 1967, joined forces with Francis Concrete of Chichester in 1981, to form Tarmac Sand & Gravel.

In late 1981, they purchased the interests of Francis Concrete, at the same time taking over its offices and its two aggregate dredgers, the *Chichester City* and the *Chichester Star*. In 1982, Tarmac registered the new company as Tarmac Sand & Gravel Ltd.

Then, in 1984, both the Bristol Channel and south coast operations were merged to form Tarmac Marine, at the same time taking over the interests of South Wales Sand & Gravel at Swansea, which had itself been formed in 1932 by the Bevan family.

In 1984 Tarmac spread their marine dredging interests into the Bristol Channel with the acquisition of South Wales Sand & Gravel of Swansea, at the same time taking over both of their ships, the *Glen Hafod* and the *Glen Gower*.

The marine dredging interests of the Hoveringham Group were taken over by Tarmac Marine in 1984, in the Bristol Channel these included Hoveringham Gravels Ltd of South Wales, included in the takeover were also ships wharves and shore plants belonging to Hoveringham.

In 1985 the *Glen Gower* was sold to Northwood Fareham Shipping Co. for service with Northwood Pioneer Concrete, working out of Fareham in Hampshire. The *Glen Hafod* and South Wales Sand & Gravel were later registered as owned by Tarmac Roadstone Holdings Ltd.

In the Bristol Channel Tarmac Roadstone was seen to operate in the same way and using the same ships as Hoveringham had done, the ships received Tarmac Roadstone colours but retained their Hoveringham names and also remained registered at Hull.

In 1987 Tarmac joined forces with Pioneer Aggregates in a joint venture, the new company resulting from this venture took the title United Marine Aggregates Ltd. The head office of this new company, was Francis House, Shopwyke Road, Chichester which was until 1981 the offices of Francis Concrete Ltd. The *Glen Hafod* was sold by Tarmac Marine in 1987 to Kendall Brothers of Portsmouth, and was renamed *KB11*.

Tarmac Marine logo.

Above: Lady of Chichester (1997), formerly *City of Chichester* and *Chichester City*,
laid up at Barry.
Below: City of Portsmouth (1994), formerly *Chichester Star*, at Cardiff.

Fleet List – Tarmac Marine

GLEN HAFOD
In fleet as a sand dredger 1984-87
History
1984 Transfers to Tarmac Roadstone Holdings
1985/86 Registered owners become Tarmac Roadstone Holdings
1987 Sold to Kendall Brothers of Portsmouth; renamed *KB II*

GLEN GOWER

Glen Gower (1984-85) passing Beachley on the River Wye.

In fleet as a sand dredger 1984-85
History:
1984 Transfers to Tarmac Roadstone Holdings
1985 Sold to Northwood Fareham Shipping; retains her name

HOVERINGHAM I

Hoveringham I (1984-87) on the River Usk, Newport.

In fleet as a sand dredger 1984-87
History:
1984 Ownership transferred to Tarmac Roadstone Holdings
1987 Transfers to United Marine Aggregates

HOVERINGHAM IV

Hoveringham IV (1984-87).

In fleet as a sand dredger 1984-87
1984 Ownership transfers to Tarmac Roadstone Holdings
1987 Transfers to United Marine Aggregates

HOVERINGHAM V

Hoveringham V (1984-87).

In fleet as a sand dredger 1984-87
1984 Ownership transfers to Tarmac Roadstone Holdings
1987 Transfers to United Marine Aggregates

United Marine Aggregates

In 1987, Tarmac and Pioneer Concrete entered into the joint venture agreement that would result in the creation of United Marine Aggregates.

Interest in having an operating base in Avonmouth dates back to 28 January 1980, when the Hoveringham Group entered into talks with the Bristol Docks Committee to open a sand berth at Avonmouth. For various reasons, this proposal was rejected. Just over ten years later, on 16 February 1990, it was announced that United Marine Aggregates was to open a one-acre site at Avonmouth Old Dock, next to the old dock entrance. It was further stated that the old lock would become a working berth, handling 200,000 tonnes of sand dredged from the Bristol Channel each year.

This berth became a joint-user berth in December 1992, when ARC Marine moved out of Bristol and set up in Avonmouth. Today, the company is known as Hanson Marine Aggregates Ltd and operates side-by-side with UMA.

During August 1993, United Marine Aggregates was in the process of updating its fleet, in particular the ship that covered the Bristol Channel area

United Marine Agreggates logo.

and Avonmouth. The company brought the *City of London* around from her base on the Thames to run some trial cargos into Avonmouth. It appears she was too large to dock at the wharf, and was unable to discharge on the tide.

Fleet List – United Marine Aggregates

CITY OF SWANSEA

City of Swansea (1987-94) in Sharpness Dry Dock.

In fleet as a sand dredger 1987-94
History:
1987 Transfers to United Marine Aggregates as *Hoveringham I*
1989 Renamed *City of Swansea* late in the year; reregistered in Swansea
1990 Damages bow in collision with quay at Briton Ferry in April
1994 Grounded at the Nash Sands in the Bristol Channel in April; sustains severe bottom damage; declared a total constructive loss; laid up in Barry; sold; taken to Cardiff for minimal repairs; renamed *Bowmore*
1996 Dry-docked at Sharpness on 29 October; sails to new owners in the Mediterranean as the *Bowmore*
1996 Renamed *Aeolos* by her Greek owners

CITY OF BRISTOL

City of Bristol (1987-97) at Swansea.

In fleet as a sand dredger 1987-97
History:
1987 Transfers to United Marine Aggregates as *Hoveringham IV*
1989 Renamed *City of Bristol* late in the year; reregistered in Bristol
1997 Sold to H. Pounds, a ship breaker in Portsmouth; towed from Swansea at 6 a.m. on 25 September
1997 Resold; renamed *L. Campo* under the Honduran flag; wrecked in position 33.19N/8.20W prior to 31 December 2008

CITY OF SOUTHAMPTON

City of Southampton (1987-97).

In fleet as a sand dredger 1987-97
History:
1987 Transfers to United Marine Aggregates as *Hoveringham V*
1989 Renamed *City of Southampton* late in the year; reregistered in Southampton
1994 Suffers a fire in the engine room while under repair in Avonmouth
1997 Sold to H. Pounds, a ship breaker in Portsmouth
1997 Sails as *Leon 1* to new owners in Greece on 12 July
1998 Renamed *Elephantus* by her Greek owners; later renamed *Kavonissi*

CITY OF CARDIFF

City of Cardiff (1997-present).

In fleet as a sand dredger 1997-present
Built in 1997 by Appledore Shipbuilders as the purpose-built motor sand dredger *City of Cardiff*
Gross tonnage 2,074 – Length 72 metres – Beam 15 metres – Max. draft 5.2 metres – Dredge depth 35 metres – Discharge rate 1,125 tonnes per hour
Machinery: Twin screw diesel engines giving 2,720 kw maximum power, twin rudders
History:
1997 Completed to the order of United Marine Aggregates as part of a two-ship order, the other ship being the *City of Chichester*; designed for service in the Bristol Channel and on the south coast, at ports suffering width and depth restrictions; capable of self-discharge by onboard crane and conveyor

CITY OF CHICHESTER

City of Chichester (1997-present), at Great Western Wharf on the River Usk.

Built in 1997 by Appledore Shipbilders as the purpose-built motor sand dredger *City of Cardiff*
Gross tonnage 2,074 – Length 72 metres – Beam 15 metres – Max. draft 5.2 metres – Dredge depth 35 metre – Discharge rate 1,125 tonnes per hour
Machinery: Twin screw diesel engines giving 2,720 kw maximum power, twin rudders
History:
1997: The same as the *City of Cardiff*, only based on the south coast

CITY OF LONDON

City of London (1993) passing Battery Point.

Built in 1989 by Appledore Shipbuilders as the purpose-built motor sand dredger *City of London*

Gross tonnage 3,660 – Length 99.74 metres – Beam 17.35 metres – Crew complement 12 – Dredge depth 36.5 metre – Loading time 3.75 hours – Discharge time <3 hours

Machinery: Twin 2,500 bhp diesel engines, by Mirlees Blackstone

History:

1989 Completed to the order of UMD as a purpose-built self-discharging aggregate dredger; part of a two ship order, the other ship being the *City of Westminster*, completed in 1990; built to serve London, the south and east coasts, and Europe

Independent Operators – The Welsh Coast

It is of interest to note how many of the aggregate dredging companies in the Bristol Channel were successively bought out to finally join and form the British Dredging Company, destined to become the largest aggregate dredging company not only in this country but also in Europe. Other companies were later formed and absorbed into dredging groups which went on to form the big three operators in the Bristol Channel. These are Cemex UK Marine, Hanson Aggregates Marine, and United Marine Aggregates.

However, some smaller companies from the early years of the Bristol Channel aggregate trade on the Welsh coast remained independent, but most of these businesses only ran for a few years before closing down. Examples of these companies are Welsh Sand and Gravel in the 1920s, Western General Shipping Company in the 1930s, and more recently Llanelli Quarries in the 1960s. Today, in 2011, Llanelli Sand Dredging, an independent operator, serves Llanelli and the surrounding area. This company is part of Westminster Dredging and has no dredgers of its own but uses tonnage from the parent company. On spring tides it carries out a dredging campaign over a two-week period to supply Westminster Dredging.

At Newport there is a very active independent operator called Severn Sands, who are managed by Cross Avon and dredge on the Bedwin Sands, near Denny Island. Their dredger supplies the company wharves at Newport, Chepstow and Briton Ferry, near Swansea. In the past, the company also dredged sea coal off Barry, which was discharged at Briton Ferry where it was screened and cleaned before being sold. This trade has now ceased and the delivering company only handles sand at Briton Ferry. On occasions, their dredger can be seen delivering a cargo of sand into Avonmouth for one of the big three operators of that port.

Welsh Sand & Gravel Co.

The Welsh Sand & Gravel Co. was one of the single-ship operators working out of the Glamorganshire Canal at Cardiff; its registered office was 59 Corporation Road, Cardiff. The company started in the sand trade in early 1920, when the owners purchased a small Glasgow-registered dry cargo coaster named *Marie*. This little ship remained unchanged during her time in the sand trade; she was one of the many early dredgers loaded by hand, and was never converted for suction dredging. Like other Cardiff-owned ships, she would have sailed out to the sand banks in Cardiff Bay, which in those days dried out between tides.

Records show that the *Marie* was built in 1891 as a single-deck vessel built with steel plates and iron frames; she remained registered in Glasgow throughout her time with Welsh Sand & Gravel. The reason for the closure of Welsh Sand & Gravel in 1926 is not known; it might be that by 1926, sand was being suction dredged. It would appear that the company's type of operation had become outdated and uneconomic. The requirement was for larger ships capable of suction dredging in six to eight fathoms of water and able to work in all types of weather.

Fleet List – Welsh Sand & Gravel Co.

MARIE (no image available)
In the sand trade 1920-26
Built in 1891 by Scott & Co., Bowling, as the steam coaster *Marie*
Gross tonnage 191 – Length 115 feet – Beam 20.5 feet
Machinery: Two cylinder steam compound, by Muir & Houston
History:
1891 Completed for McKinney & Rafferty; registered in Glasgow
1904 Sold to Kynoch Arklow
1920 Sold to Welsh Sand & Gravel
1926 Sold to S. Gray of Belfast; reregistered in Belfast

Western General Shipping Co.

Western General Shipping Co. was involved in the coastal trade and worked out of the docks at Cardiff and Barry, with a registered office at 34 Romilly Park, Barry, Glamorgan. In 1929-30, the company was listed in the Lloyds Register of Shipping under the name 'T. White' as a one-ship company owning the coastal cargo vessel *Cape Wrath*.

Late in 1929 or early in 1930, the company purchased the ex-Admiralty steam coaster/supply vessel *Sir Redvers Buller*. She was renamed *Redvers*

Buller and was used by the company in the Irish trade. In 1931, she was converted in Cardiff into a sand suction dredger and was put to work in the South Wales sand trade. At the same time, the company was renamed the Western General Shipping Co. and was based at Cardiff with T. White as its manager. The company's original ship, the *Cape Wrath*, remained in the coastal trade; she was never involved or used in the sand trade.

Western General Shipping's venture into the sand trade was to be short-lived and came to an end on 14 October 1932. The *Redvers Buller* loaded a cargo of sand at the Holms and was bound for Swansea when she took a sudden list in the early hours of the morning while at anchor. Without warning, she capsized and sank, with the loss of four of her crew. *Redvers Buller* became a total loss, after which Western General Shipping reverted to the coastal trade using the *Cape Wrath*.

Fleet List – Western General Shipping Co.

REDVERS BULLER
In fleet as a sand dredger 1931-32
Built in 1895 by Cook Welton & Gemmel as the admiralty coaster/supply vessel *Sir Redvers Buller*
Gross tonnage 266 – Length 130.7 feet – Beam 23 feet
Machinery: Steam compound two cylinder. 62 bhp, by Amos & Smith, Hull
History:
1895 Completed for the British Admiralty as a supply vessel
1929 Sold to T. White of Cardiff; renamed *Redvers Buller*; reregistered in Cardiff
1931 Converted at Cardiff into a sand suction dredger
1931 Owners become Western General Shipping Co.
1932 Capsizes and sinks after loading on 14 October; lost, with four crew

Severn Sands

Severn Sands Limited of Newport was first established in the sand trade using hired tonnage. It is understood the company hired the Swansea-owned *Glen Gower* to run into Newport and, when required, up the River Wye to Chepstow. Then in 1985 they purchased the sand dredger *Rhone* from Bury Sand at Swansea, which by this time was part of the RMC group.

The company to establish an aggregate trade at Newport and Briton Ferry used the *Rhone*; they also used her to dredge sea coal off Barry, which was discharged at their Briton Ferry wharf. In 1995 the *Rhone* was replaced by the French-owned sand dredger *Ferlas*, a ship well known at Newport and in the Bristol Channel as the Newport-owned *Isca*. The company renamed her *Severn Sands*. The *Rhone* was withdrawn from service and laid up at Newport, where she was offered for sale and later sold to foreign owners. The company discontinued their sea coal trade to Briton Ferry, but continued to trade with sand there, Newport and Chepstow. Further expansion took place in 2004 when the company purchased the Yorkshire-class coaster *Hoo Maple*, which they then converted into a self-discharging sand dredger, which they named *Argabay*. With the delivery of the new dredger, the company was not only able to supply their own wharves, but could also do one of cargoes for other companies in the Bristol Channel. The *Severn Sands* was laid up firstly at Newport and then at Chepstow before being sold to North Devon owners.

Fleet List – Severn Sands

RHONE

Rhone (1985-95).

In fleet as a sand dredger 1985-95
History:
1985 Purchased by R. Brean of Newport from Bury Sand of Swansea
1996 Sold to owners in Funchal; renamed *Ribiera Grande*

SEVERN SANDS

Severn Sands (1995-2005) at Newport.

In fleet as a sand dredger 1995-2005
History:
1995 Purchased by Severn Sands from owners in France; returns to Newport for repairs; renamed *Severn Sands*
2005 Makes her final commercial voyage on 28 April; laid up; offered for sale
2005-07 Laid up in Newport then Chepstow
2007 Sold to owners in North Devon in May
2009 Breaks free from moorings at Fremington Quay; considered to be a hazard; awaiting scrapping
2010 Scrapped at Yelland in November

ARGABAY

Argabay (2005-present).

In fleet as a sand dredger 2005-present
Built in 1984 by Yorkshire Drydock Co. as the motor coaster *Hoo Maple*
Gross tonnage 756 – Length 58.3 metres – Beam 9.4 metres
Machinery: Twin Aqua master propellers, twin Crossley diesel engines
History:
1984 Completed for Whitaker Holdings Ltd with R. Lapthorn as managers
2000 Ownership transferred to R. Lapthorn & Co.
2004 Sold to R. Brean of Crossavon Ltd, Newport
2004 Drydocked at Newport in June; renamed *Argabay*; reregistered in Newport
2004/05 Converted at Barry into a sand dredger by Harris Pye Marine
2005 Enters service as a sand dredger on 4 May

Llanelli Quarries

Llanelli Quarries purchased the *Catherine Ethel* from Cardiff, intending to use her at Llanelli, but for various reasons she was unsuitable. She ran only four or five cargos, then was laid up. An attempt was later made to convert her into a diesel vessel, but this was abandoned and after a further lay up she was scrapped in Llanelli. Nothing further is known about this company. It was said that they pulled out of sand dredging and reverted to quarrying operations.

Fleet List – Llanelli Quarries

CATHERINE ETHEL

Catherine Ethel (1962-69) at Cardiff.

In fleet as a sand dredger 1962-64
History:
1962 Purchased from Sandridge, Cardiff
1963 Found to be unsuitable for operating in Llanelli, due to a lack of good-quality steam coal
1964 Work on diesel conversion commences; ceases soon afterwards; laid up in Llanelli
1969 Broken up in Llanelli

Llanelli Sand Dredging

Llanelli Sand Dredging, a subsidiary of Westminster Dredging (in turn part of the Royal Boskalis Westminster Group), was established in 1993 to serve the local area from its base at Burry Port. Unlike other sand dredging concerns in the Bristol Channel, Llanelli Sand Dredging operate two dredging campaigns a year, which each last for four to five weeks; this allows the company to stockpile the distribution site. Dredging is undertaken at the Hellwick grounds by the dredgers *Sospan* and *Sospan Dau* ('Sospan Two'). Both ships discharge at a pontoon in the Loughor Estuary, from where the cargo is pumped ashore to a stockpile in a nearby field. When the ships are not dredging sand for the company, they are involved in harbour and dock maintenance work for the parent company.

Fleet List – Llanelli Sand Dredging

SOSPAN

Sospan (1993-present) at Briton Ferry.

In fleet as a sand dredger 1993-present
Built in 1990 at Papendrecht as the purpose built motor dredger *Sospan*
Gross tonnage 718 – Length 57 metres – Beam 10 metres
Machinery: Main diesel engine, 260 kW bow thrust unit
History:
1990 Completed for PVW Skua; owners later become Westminster Dredging

SOSPAN DAU

Sospan Dau (2001-present) at Burry Port.

In fleet as a sand dredger 2001-present
Built in 1978 at Atelieres, Dieppe, as a motor tank barge, converted into a dredger in 2001 by Kooiman BV, Holland
Gross tonnage 1,850 – Length 71 metres – Beam 14.3 metres
Machinery: Main diesel engine, 283 kW bow thrust unit
History:
1978 Completed as a motor tanker
2001 Purchased by Westminster Dredging and converted in Holland into a motor dredger

CHAPTER 2

Surviving Welsh Sand Dredgers

Severn Sands

North Devon is the resting place of a Welsh Bristol Channel sand dredger. This vessel, the *Severn Sands*, is currently (as of January 2010) beached at Penhill, near Fremington Quay. She was built in 1960 by NV Scheeps at Westerbroek in Holland, to the order of Western Dredgers of Newport. Named the *Isca*, she was a purpose-built sand dredger, built like a Dutch motor coaster. She was a very useful size for the Bristol Channel aggregate trade, being 550 gross tons on a length of 51.67 metres. Initially an independent company, Western Dredgers were by the early 1970s working closely with British Dredging, having found it difficult to compete. In 1976, Western Dredgers became wholly owned by them, and the *Isca* was transferred to their ownership.

Offered for sale by British Dredging Company in 1977, the *Isca* was sold to Louis M. Periou of Pampol, France who re-named her *Le Ferlas;* she kept this name until 1989, when she took the shortened name of *Ferlas*. She had worked continuously in French waters for almost twenty years, and in 1995 she was offered for sale, although at thirty-five years of age, many thought she might go to the breaker's yard.

Purchased by Severn Sands of Newport for further trading as a sand dredger, she too returned to her original homeport, arriving at Newport on 5 April 1995. For many it was good to see her return, but her bright green hull was something of a shock. She was dry-docked for survey, repainted, renamed *Severn Sands*, and registered in Newport. Working for Severn Sands for almost ten years, she ran her last cargo to Chepstow on 28 April 2005, after which she was initially laid up at Newport and later at Chepstow.

A buyer was found in North Devon, where, under her own power, she arrived on March 2007 and was moored on the old oil jetty at Yelland, before being moored at Fremington Quay. On her arrival at Fremington, she was to become a tourist attraction and her sand tank was to be converted into a steam museum; later it was said she was to be converted into an eco-friendly home. She was stripped of much of her equipment, including anchors and the propeller, in Fremington.

Severn Sands, formerly the *Isca*, here the *Ferlas* in France (1980s).

During a gale in March 2008, she broke adrift and at one point threatened to block the river. Luckily, as the tide turned, the wind dropped and she was grounded at Penhill, a small beach above Fremington, where she now lies. Since that time, she has been mysteriously filled with old tyres, cans of paint, and other unknown substances, and could be described as an environmental time bomb. There have been difficulties with tracing her owner; the local authority had put aside money for her disposal. Late in 2009, the owner came forward, stating his intention to re-moor the *Severn Sands* somewhere in North Devon, and to convert her into an upmarket night club.

The author has happy memories of this ship, having witnessed her arrival from the builders on a foggy day in 1960; her grey-painted hull, favoured by Dutch shipbuilders, was immaculate. It was sad to see her in the summer of 2009, and in many ways it would have been better had she ended her days in the breaker's yard.

Kyles

Moving to Scottish waters and to Braehead, we find the oldest survivor of the Bristol Channel sand trade, the *Kyles,* dating from 1872. She has been restored to her 1953 appearance and is now a floating, working exhibit at the Scottish Maritime Museum. Her story starts in March 1872, when she was launched from Yard No. 11 of the Paisley shipyard of John Fullerton & Co. She was built as an iron-hulled steam coaster of eighty gross tons on a length of 82.5 feet, and was powered by a two-cylinder twenty horsepower steam engine – built by King & Son of Glasgow – which drove a single screw.

Her first owner was Stuart Manford of Glasgow, who used her as a support vessel for the fishing fleet, taking stores out to the trawlers and returning

Above: Kyles sand dredging in the River Avon (1921-42).

Right: Advertisement for Leyshon Steamers showing the *Kyles*.

with fish for the markets. She had nine owners before arriving in the Bristol Channel in 1901, when she was owned by John Leyshon and used as a goods packet boat between Bristol and Cardiff. A further four owners in the Bristol Channel followed, until she was sold in 1919 to owners in Herne Bay.

In 1920, she returned to the Bristol Channel. In 1921, she was sold to the Cardiff tug master George Hamlin, who had her converted into a sand dredger. He worked her out of the Glamorganshire Canal in Cardiff.

At the start of the Second World War, she was laid up in the canal and in 1942 she was surveyed and found to be in poor condition. She was then sold to William Metcalf of Ilfracombe as a salvage craft.

Kyles as a sludge tanker at Sharpness (1960s).

It seems doubtful that she was ever actually used by William Metcalf. She spent her time laid up in Ilfracombe Harbour. In 1947, she was sold to Ivor Price Langford of Sharpness, who in 1953 rebuilt her as a motor coaster. She was fitted with a four-cylinder 120-bhp diesel engine, making her 121 gross tons. Further changes were made during the ownership of Ivor Langford. In 1960, she was again converted, this time to a sludge tanker. At the end of her time in his ownership, she became a storage hulk.

In 1981, she was again rescued, this time by Peter Herbert of Bude, who purchased her as a restoration project; he took her to the basin at Bude Canal, but the project came to nothing. Peter Herbert sold the *Kyles* to the Scottish Maritime Museum in 1984, and single-handedly sailed her from Bude to Irvine. In 1996, funding became available for her full restoration. It was decided to restore her to her 1953 appearance, as it would be impractical to return her to her original condition. The restoration work was completed in 1999, when the *Kyles* ran her sea trials.

She is a remarkable little ship. In her lifetime of 138 years, she has kept the same name, but had twenty-four owners. She has been modified three times and rebuilt twice. Of her 138 years, she spent seventy-eight years in the Bristol Channel – fifteen years between 1901 and 1916 and then a further sixty-three years between 1921 and 1984. For twenty-one years of this, she was a sand dredger.

Good fortune has followed her since 1942, when William Metcalf first rescued her. She was again rescued by Ivor Price Langford, who is known to have considered *Kyles* his favourite ship. She was the only one he never renamed. She was yet again rescued by Peter Herbert of Bude, and then finally by the Scottish Maritime Museum. This is the remarkable story of a true survivor in the world of shipping.

No. *2435*   Date *3ʳᵈ November 1921*

Name and Description, *Steel Screw Sand Suction Dredger "Kyles".*

Date of Launch, *1918*   Port of Registry, *Hull*   Off. No. *54706*

Name and Address of Owners' *Sandridge & Coy, Hugh Hawkin, Manager, Cardiff*

Name and Address of Builders, *John Fullerton & Coy, Paisley*

Registered Dimensions, *80.9' × 18.15' × 4.4'*   Moulded Dimensions, *83'0' × 18'1' × 8'1'*

Registered Tonnages ; Gross, *49.96*   Nett, *23.8*   Under Deck,

Freeboard in Summer, from Statutory deck } *1'0"* Allowance in Fresh Water, *3"* Winter, *1½"* Winter North Atlantic, - -

line *1* ins. above ~~iron~~ wood deck at side.}

5ᵗʰ Special Survey No. 1 is due October 1925

PARTICULARS OF MACHINERY.

Engines, Description, *Compound*

Name of Maker, where and when built, *Wm King & Coy Glasgow 1918*

Diameters of Cylinders, *16" and 23"*   Length of Stroke, *18"*   Cub. ft. L. P. Cyl. *11·5*

Boilers, Description, *Multitubular*   Forced Draught, description,

Number, Single ended, *1* No. of Furnaces in each, *2* Double ended, — No. of Furnaces in each,

Name of Maker, where and when made,

Dimensions, ᴱˣᵀᴱᴿᴺᴬᴸ mean diameter, *8'11"*   Length, *9'1"*

Heating Surface,   Grate Surface, *26 Sqr Feet*   Working Pressure, *95 lbs*

5ᵗʰ Special Survey No. 1 is due October 1925

(SIGNED) *Geo. K Chamberlain for Chairman.*

(SIGNED) *J. L. Adam for Chief Surveyor.*

(SIGNED) *John Cuming Secretary.*

Date of Reporting, *22ⁿᵈ October 1921 at Cardiff.*

CLASS B.S. *(Suction Dredger).*

Copy Survey Certificate issued 3 November 1921 for Suction Dredger *Kyles* when she was owned by Sandridge & Co.

CHAPTER 3

Welsh Archaeology

Moderator Wharf

Moderator Wharf was a wharf I often visited when away on the dredgers. Close to the town bridge, there were three sand wharves, one of which was Moderator Wharf. When the sand trade moved downriver, Newport Council decided to redevelop these wharves, and Moderator Wharf was to become a new arts centre.

It was during the excavations for the basement of this building in the summer of 2002 that the remains of a well-preserved medieval ship were found in the mud. It later became known as the Welsh 'Mary Rose', and is said to be the finest example of a fifteenth-century ship. Archaeologists also found remnants of clothing, pieces of pottery, stone cannon balls, and lumps of cork. Some of this suggests that the ship was trading to Portugal.

The mud at Moderator Wharf.

Left: 'The Newport Ship', found at Moderator Wharf.

Below: A closer look at the hull. More ribs are exposed.

Isca on the River Usk at Moderator Wharf, Newport (1997).

The ship itself, the hull of which is almost intact, was over twenty-five metres in length and is thought to be a cross between a Viking long ship and a caravel. Newport City Council and the Welsh Assembly set aside funds to preserve the ship prior to it being put on display. It is thought in excess of 1,700 timbers were used in her construction. After surveying, measuring, and photographing the ship, the remains were carefully removed and placed in tanks of water to preserve them until a suitable site is found to exhibit her.

When the Newport ship was removed in December 2002, the remains of a skeleton were found beneath her. It was first thought that the skeleton was connected with the ship, but this theory was proved incorrect after tests on the bones have shown them to date back to 170 BC, about 1,500 years before the ship was built. Further tests have shown the skeleton to be of a very muscular male about 5 ft 9 in in height, probably in his late twenties or early thirties.

CHAPTER 4

Ports of Call
and Operational Bases

Newport Fleets

Newport Sand & Gravel (1923-63) operated in Newport, Cardiff, and Swansea.

Western Dredgers (1960-76) initially served Newport, Cardiff, and Swansea, later the whole of the Bristol Channel as a part of British Dredging.

Severn Sands (1985-present) served Newport, Chepstow, and Briton Ferry. The company also runs to the wharves of other operators in the Bristol Channel.

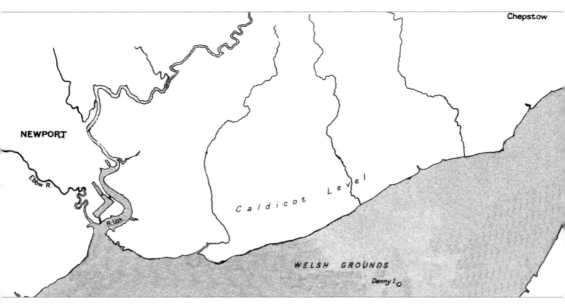

Newport and Chepstow.

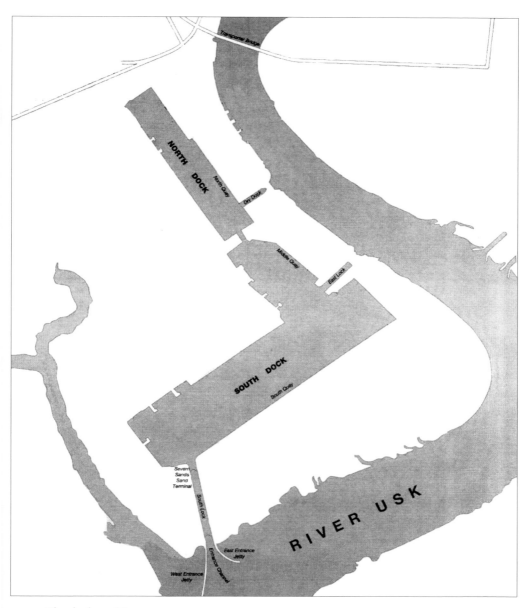

The docks at Newport.

Moderator Wharf, Newport (*c.* 2009).

Rhone at Newport.

Instow.

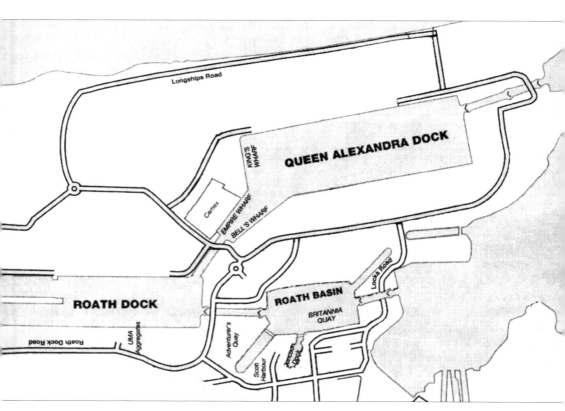

Cardiff Docks (2009).

Minehead Harbour in the 1930s, showing Arthur Sessions ship *Corbeil* discharging.

Cardiff Fleets

Arthur Sessions & Son (1914-65) worked in Newport, Cardiff, and Swansea. In the early years, the company also ran to Gloucester, Bristol, and Watchet.

Welsh Sand & Gravel Co. was a one-man and one-ship operation based in Cardiff.

William J. Holway was a one-man and one-ship operation based in Cardiff.

Western General Shipping was a single-ship operation based in Cardiff. George J. Binding, Cardiff, was a single-ship operation based in Cardiff.

F. Bowles & Sons (1920-62) operated in Newport, Cardiff, Barry, Briton Ferry, and Swansea.

Sandridge & Co. (1921-62) operated in Cardiff. Its ships could be also seen in Newport. In the early years, the company also ran to Bristol.

Davies, Middleton & Davies (1962-63) and Sand & Gravel Marketing (1962-65) were based in Cardiff and served Newport, Cardiff, and Swansea.

British Dredging, later British Dredging Aggregates (1962-90), covered the whole of the Bristol Channel except North Devon.

Hoveringham Gravels (1965-84) served Newport, Cardiff, and Swansea.

Barry Docks (*c.* 1986). *Peterston* is in the No. 3 Dock.

Bowcross discharging at Queen Alexandra Dock, Cardiff (1995).

Hoveringham I at Cardiff.

City of Portsmouth at Briton Ferry.

Swansea Fleets

South Wales Sand & Gravel (1932-84) served Swansea and Briton Ferry. Later, its ships were seen in Sharpness and Chepstow.

Channel Sand & Ballast (1948-63) mainly worked out of Swansea. The company's ships were also seen working in Cardiff and Newport

Tarmac Roadstone Holdings, later Tarmac Marine (1984-87), was based in Swansea but worked to other South Wales ports.

Bury Sand (1982-90) operated in Swansea.

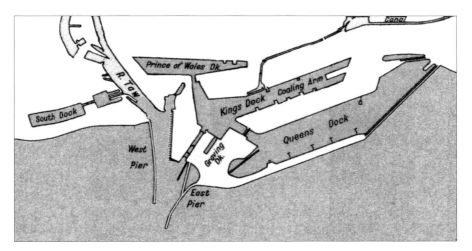

Swansea Docks.

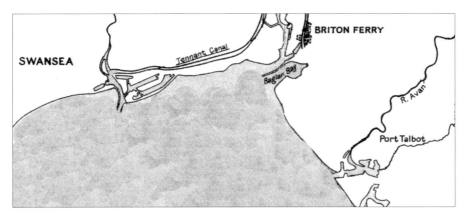

Swansea and Port Talbot.

Glen Gower and *Sand Jade*, River Usk (1976).

Llanelli Fleets

Burry Sand (1965-78) operated in Llanelli.

   Llanelli Plant Hire (1978-82) operated in Llanelli and Swansea.

   Llanelli Sand Dredging (1992-present) operated in Llanelli, with a discharge berth in the Loughor Estuary.

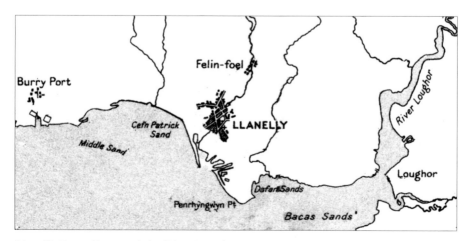

Llanelli, Burry Port, and the River Loughor.

Sospan Dau in the Loughor Estuary.

CHAPTER 5

Life on Sand Dredgers

Dangers Faced by Crews

Like any other maritime occupation, aggregate dredging has its own share of dangers. On one hand, there are constraints set by time and weather conditions, such as tidal deadlines, fog, and gale force winds. On the other, there is the possibility of mechanical or structural failure within the ship, sometimes caused by external forces. Other dangers include collision. Due to their cargo being almost liquid, sand dredgers are notoriously unstable.

The Bristol-owned sand dredgers *Dunkerton*, of Bristol Sand & Gravel, and *Sandholm*, of The Holms Sand & Gravel, had sailed from Bristol in good weather conditions on 17 October 1960 bound for the Holms dredging grounds in the Bristol Channel to load. Nearing the Horse Shoe Bend in the River Avon, both vessels were caught in blanket fog. The *Dunkerton* hit the rocks at the Powder House, badly damaging her starboard bow, but managed to reach Avonmouth. The *Sand Holm* ran ashore in the river and settled at an alarming angle, damaging her rudder in the process; she was safely refloated on the next tide. Both ships later returned to Bristol for inspection and repair.

The Barnstaple-owned *Stan Woolaway* of William Woolaway & Sons was returning to Barnstaple, having loaded a full cargo at the Holms on 13 March 1967. Near Bull Point in North Devon, she developed a sudden list and sank shortly afterwards. Her loss was caused by the fact that the cargo could not be drained due to a collapse of her drainage system, which made her unstable.

On 4 October 1968, the Bristol-owned *Steep Holm* of The Holms Sand & Gravel, having dredged a cargo of sand off Barry, was bound for Swansea fully laden. On the voyage down-channel from the dredging ground, the weather started to deteriorate – high wind and heavy seas. The *Steep Holm* was driven ashore on the Tuskar Rock near Porthcawl; her crew were rescued by the Mumbles Lifeboat and landed safely. The *Steep Holm*, battered by the high winds and heavy seas, started to break up almost immediately and became total loss. One salvage worker lost his life.

The *Steep Holm* on Tusker
Rock, near Porthcawl.

On 20 October 1989, the Bristol-owned *Sand Diamond* of Sand Supplies
(Western) was at the dredging grounds near the Holms in southerly winds
of force six/seven when a mechanical breakdown disabled the ship's pumps,
causing her to take on water and list. A Mayday call for assistance was sent
out, and a helicopter landed an auxiliary pump on her deck. By this time, it
was estimated that she had shipped 100 tonnes of water. Later, with the aid of
this pump, the *Sand Diamond* was able to return to Bristol for inspection and
repair.

The Bristol-owned *Sand Sapphire* of Sand Supplies (Western) was dredging
in July 1980 when a live mustard gas bomb from the First World War was
dredged up onto the deck. The crew radioed for help and waited until the
bomb disposal squad arrived to deal with their unexpected guest. This time,
the crew and the ship were lucky!

The Southampton-owned *Sand Skua* of South Coast Shipping was dredging
at the Solent Bank on 30 May 1972 when there was an apparent blockage in
the dredge pump. Close inspection showed that the pump room was rapidly
filling with sea water. It was later found that a sixty-five-pound artillery
shell had been dredged up, and had exploded in the dredge pump. With this
explosion, the pump had shattered, opening the pump room to the sea. The
Sand Skua sank, bow down and with a list to port. She had cost £500,000 to
build and her salvage was estimated to have cost around £125,000, with a
further £50,000 in repair costs.

Above: The *Sand Skua* sinking in the Solent.
Below: Another view of the *Sand Skua*, *c.* 1972.

Above: The *Sand Skua*.
Below: *Sand Skua* as a cement carrier at Sharpness, *c.* 1988.

Sand Skua rebuilt as sand dredger *Norstone*, based at Fareham and working on the South Coast. She was owned by Northwood Fareham Shipping, part of Westminster Gravels.

Aggregate Dredgers in Wartime

The three dredgers owned by Bristol Sand & Gravel were kept very busy during the First World War. From Bristol and the ports of South Wales, tens of thousands of tons of sand were supplied to ballast war transport ships involved in ferrying troops and munitions to France.

The sand dredgers were again very busy during the Second World War, supplying sand for the ballasting of war transport ships and the production of sand bags. The company also supplied aggregates for the building of airfields and the construction of factories. These factories built aircraft, manufactured munitions, and all manner of war goods. It should be noted that the dredger fleets in South Wales were also employed in the war effort, though in the Second World War only, as the South Wales trade developed after 1920.

In 1941, the Bristol Sand & Gravel dredger *Dunkerton*, berthed at the Welsh Back in Bristol, was caught in a heavy air raid on the city. She sustained severe damage but remained afloat. Immediate repairs were carried out to put her back in service. She was again attacked in the Bristol Channel, but only sustained minor damage, with no crew injuries. *Dunkerton* carried the signs of shrapnel damage on her superstructure; there was even a piece of shrapnel on a mounted teak wood block in her saloon. It was said her captain kept it as a reminder of the 1941 attack on his ship.

Durdham.

Not so lucky was the Bristol Sand & Gravel dredger *Durdham*, which was on charter to British Dredging in Cardiff. She had sailed from Cardiff on 27 July 1940 bound for the Holms, to load another cargo for Cardiff. When outward bound at Lavernock on the Welsh coast, she struck a magnetic mine and sank with all hands. Her six-man crew are remembered on panel number thirty-six of the Merchant Navy Memorial at Tower Hill.

Another Bristol Channel sand dredger lost to enemy action was the Cardiff-owned *Skarv*. At the time of her loss, she was on charter to the Admiralty. She had sailed from Swansea on 11 November 1940 bound to the Nash Sands to load another cargo; her orders were then to return to Swansea and discharge. Nothing more was heard from the *Skarv*, and she was posted as missing and untraced. After the war, she was considered to be a war loss. The probable cause was that she struck a mine. Her five-man crew are remembered on panel number ninety-eight of the Tower Hill memorial.

The wording on the Memorial reads:

1939-1945
THE TWENTY-FOUR THOUSAND OF THE MERCHANT NAVY
AND FISHING FLEETS
WHOSE NAMES ARE HONOURED ON THE WALLS OF THIS GARDEN
GAVE THEIR LIVES FOR THEIR COUNTRY
AND HAVE NO GRAVE BUT THE SEA

The Merchant Navy War Memorial at Tower Hill, London.

Another view of the Merchant Navy War Memorial.

The memorial on Welsh Back, Bristol, to the
men of the Merchant Navy and fishing fleets.

The Merchant Navy Association War Memorial at the Welsh Back in Bristol
City Dock is in the form of a capstan with seats around a paved area; on the
seats are brass plaques bearing the names of the seamen who appear on the
roll of honour. The Memorial bears the following inscription:

MERCHANT NAVY ASSOCIATION BRISTOL
THIS MEMORIAL IS A TRIBUTE TO ALL SEAFARERS WHO
SAILED TO AND FROM BRISTOL THROUGHOUT THE AGES
UNVEILED BY HRH THE PRINCESS ROYAL
21 MAY 2001
ETERNAL FATHER STRONG TO SAVE

The Cardiff Bay Merchant Seafarers War Memorial was built in 1997, and
commemorates the Merchant Seamen of Cardiff, Barry and Penarth who left
during the Second World War never to return.

The sculpture is unique in that it has two aspects: the face of a person called
the 'Timeless Face' and the hull of a beached ship.

Set in the new Cardiff Bay Development and facing across the bay to the
Bristol Channel it is a most fitting tribute to the seamen who lost their lives.

The Merchant Seafarers' War
Memorial in Cardiff Bay.

Interviews

Some years ago I was fortunate enough to speak with a number of the masters of the Bristol Channel aggregate dredgers, who related significant incidents they had experienced with their ships.

One such event was described by the captain of the *Badminton* who, in February 1963, was bound for Bridgwater with a cargo of road grit dredged from the Bristol Channel. This grit was to be used on icy roads and not as surface dressing. The vessel entered the river at Burnham-on-Sea to find it was frozen over. Proceeding upriver, the *Badminton* missed the tide and was grounded on pack ice; when she floated off on the next tide, she was discharged and returned to Bristol for inspection. This inspection showed that she had badly damaged her bottom plates, and she was immediately sent to Troon for repairs. A whole new bottom was fitted. The captain told me that this whole incident was caused by time and tide. The time taken to navigate the river increased dramatically with the ice, causing him to miss the tide on the way to the berth.

A similar incident happened to the captain of the *Bowcrest* when he left Cardiff one day, with orders to load at the Nash dredging ground and proceed to Swansea. Having sailed from Cardiff late on the tide and then arrived at the Nash, he proceeded to load a full cargo of sand for Swansea. This done, and the dredge gear stowed, he picked up anchor, only to find he had grounded on the Nash bank. Much to his embarrassment, he was high and dry, with other Swansea-bound vessels passing him. He was on top of the bank he had been dredging, whereas the navigation channel was between the sand bank and the Nash cliffs.

The mate of the *Bowstar* explained to me how in 1954 the captain had overcome a serious problem. Having loaded at the Holms dredging ground, the *Bowstar* took a sudden and dangerous list after her cargo shifted. Fearing the ship was in danger of capsizing, her captain took the decision to sail the vessel to Newport for discharge. This was an amazing feat, since he sailed her the entire way stern first, and in so doing, prevented further problems. After a long and eventful voyage of over six hours, the *Bowstar* was berthed safely at Newport.

A former crew member of the *Steep Holm* explained to me the severity of conditions in January 1963. A dramatic drop in temperature caused the Bristol City Dock and the River Avon to freeze over. He also said that coming up-channel from the Holms, they encountered small icebergs, which had been caused by snow clearance from the roads being dumped in the channel. On arrival at Bristol, the *Steep Holm* was breaking sheet ice to enter the lock. The situation was almost as bad in the dock, but movement of shipping had cleared a pathway. However, on arrival at the entrance to Bathurst Basin and the *Steep Holm's* berth, they had to keep going ahead and astern on the engine for the ship to break way through the now-thick ice.

Another example of time and tide described to me concerned the *Sand Diamond* in 1986, which had entered the River Avon late on the tide, fully loaded and bound for the company berth at Hotwells. She got up river as far as Hotwells pontoon, where she touched bottom and grounded; she was refloated on the next tide, unloaded, and later inspected for damage on the grid iron. This shows how dangerous the River Avon can be. Ships in similar circumstances have ended up with severe damage or even a broken back.

On 8 May 1989, the Bristol dredger *Harry Brown* was on the River Avon, fully loaded and bound for her berth at Hotwells. She was involved in a collision with the Wessex Water Authority's sludge vessel *Glen Avon*. The bow of the *Harry Brown* was badly damaged and the *Glen Avon* received severe damage to her port side and superstructure. It was said the collision was part-caused by protestors in a rubber dinghy, who had been circling near the sludge berth at Shirehampton to object to the disposal of sewage at sea. They were attempting to prevent the sludge vessel *Glen Avon* from sailing, and this caused the *Glen Avon* to leave her berth for safety reasons. This put the two ships into each other's path, and with little or no time for avoidance, the two ships collided. Luckily, there were no injuries on either ship, but both ships were out of service for some time, while repairs were effected.

Another story took place on 26 November 1946 and concerned the loss of the Bristol-owned dredger *Garth*. She was on charter to British Dredging at Cardiff and having completed her week's work was ordered to load a cargo at the Holms dredging ground, to be discharged at the berth of Bristol Sand & Gravel. It was a clear and fine night as the *Garth* proceeded up channel to Bristol. At Walton Bay, ships were anchored, waiting to berth at Avonmouth. The *Garth* ran over the out-stretched anchor chain of the unlit *Blair Devon*, holing her underwater and causing her to flood immediately. Her captain attempted to beach her in Walton Bay, but she sank soon afterwards and was a total loss.

The Barnstaple-owned *Ron Woolaway* was anchored at the Holms on 18 June 1960; the weather was calm but there was a dense channel fog. Without warning, the *Ron Woolaway* listed heavily and rolled over, floating bottom side up. Her dredge pipe was still stowed, and dredging had not commenced. Purchased only a short time before in Holland as a motor coaster and converted for sand dredging, her capsizing was blamed on the fact that she was too long, narrow, and high in design. She was later repaired and modified by lowering the overall height of the superstructure and adding tanks, or blisters, to the sides of her hull and so increasing her beam and improving stability.

One very great danger faced by dredger crews is the loss of stability before and during loading, when the dredger is de-ballasted and loaded with an almost liquid cargo. This happened in the case of the Cardiff-owned *Bowqueen*, which on the 8 September 1965 was dredging in gale-force conditions. She capsized and sank off Walton-on-the-Naze, with the loss of four of her crew. She was later raised, repaired, and put back into service. Many of her later crews believed she was haunted; in bad weather, they often thought they heard screaming from below deck, and believed these screams came from the crew trapped when the *Bowqueen* sank.

Another stability incident was that of the Cardiff dredger *Redvers Buller* in October 1932. She had sailed from Swansea and was anchored at the Holms to load a cargo. Having completed loading, she suddenly took a severe list and sank in less than three minutes. Of her seven-man crew, only three survived: her captain, the mate, and the chief engineer. This was a particularly sad story, as the owners of the ship had bought her during the depression to give out-of-work seamen a chance of employment in South Wales.

An Interesting Collision

The sand trade at Cardiff had started in the Glamorganshire Canal, used by the small sand dredger companies like Arthur Sessions, F. Bowles & Sons and Hugh Hamlin, all of whom used the canal as their base. Later, with larger tonnage ships being used, they moved to other wharves in the West and East Bute docks in Cardiff; however Bowles set up a wharf at an area known as the Ferry. The small vessels remained in the canal, which was owned by Cardiff Corporation, who wanted to close it to commercial traffic due to high running costs.

Late on the night of 5 December 1951, the sand dredger *Catherine Ethel*, owned by Hugh Hamlin of Cardiff, collided with the lock gates at the sea lock. The lock gates were destroyed, causing the canal to drain over a mile long section. The lock gates were never repaired ending 158 years of operation of the canal, and allowing the City Corporation to close the canal to commercial traffic.

CHAPTER 6

Dredgings

Facts, Reminiscences, Tales, and Trivia

Cement was a major commodity for Bristol Sand & Gravel. The company had a cement shed at its Dundas Wharf site in Redcliffe Street. This was fed from a large storage shed on the Welsh Back. Cargos of cement were imported using coastal vessels arriving from Europe, Ireland, and London. A large trade in cement had been built up and the company found it more profitable to hire coasters for cement cargos, and not run their own cement ships.

'Running the tide' was a phrase used when the dredger sailed on the first of the tide down to the Firefly Buoy, loaded a part cargo (normally 350-400 tons of sand), then returned to Bristol on the last of the same tide. 'First of the tide' was the first possible time to sail or dock, 'last of the tide' was the last possible opportunity to sail or dock on the same tide.

'Turning around' was where the loaded dredger arrived on the first of the tide – normally in the River Usk at Newport – discharged its cargo, and sailed on the last of the same tide, in some cases with 100-150 tons still on board.

In July 1956, the Glasgow-registered steamship *Yewcroft*, bound from London with a cargo of cement for Bristol Sand & Gravel, ran ashore in fog and broke her back on rocks at Cudden Point, Mounts Bay, Cornwall.

The *Ardri* at Welsh Back, having discharged a cargo of cement. Ashmead's tugs *Hubert* and *Benfleet* are in the foreground.

A delivery of fresh, clean seawater for Bristol Zoo.

The dredger *Camerton* was built with a storage tank, which was filled with clean seawater during loading. This seawater was sold to Bristol Zoo for use in the penguin, sea lion, and polar bear enclosures. As the zoo needed clean seawater, the water was normally loaded down-channel, where it was clean and not muddy.

In 1953, a large section of the tusk of a mammoth was excavated at the Hoveringham quarry site, near Nottingham, after which a woolly mammoth was adopted as the company colours, intended to show the strength of the company in business.

Llanelli Sand Dredging's ships were named *Sospan* and *Sospan Dau* in honour of Llanelli and the tinplate works, which supplied tinplate for the manufacture of saucepans. For this reason, Llanelli earned the nick name 'Tre Sospan' ('saucepan town').

Some time ago, information was passed to me about T. R. Brown's expansion out of the Bristol Channel. The story goes that a member of the Brown family went on holiday to seaside locations with a suitcase and jam jars to collect samples of sand. Back home in Bristol, these samples were analysed in order to see where the best supply of good-quality sand could be found. The Brown family then made a decision about where they would expand their dredging interests.

Later, it was not unusual for a dredger to be sent out for a week or two prospecting new dredging grounds. After its formation, British Dredging did this with the *Bowcrest*.

When the *Peterston* sailed on 13 November 1975, Dundas Wharf finally closed to shipping, and the sand dredgers transferred to Pooles Wharf at

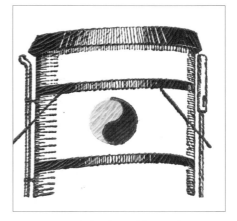

The funnel of *Hoveringham I*. Notice the black mammoth.

A drawing of the funnel logo of Bristol sand & Gravel, taken from the company's souvenir brochure.

Hotwells. Shortly afterwards, the large Stothert & Pitt crane at the Dundas Wharf was dismantled and shipped down to Hotwells, where it was rebuilt to serve the new sand wharf.

Bristol Sand & Gravel displayed its company logo, 'Pioneers of Sand Dredging', on the doors of its lorry fleet and on the back of the crane at Dundas Wharf – which could be seen from Redcliffe Street. The logo consisted of a red comma and a blue comma entwined to form a ball, around which were the words 'Pioneers of Sand Dredging'. This logo, without the wording, was also carried on the funnels of the company's sand dredgers.

In 1956, when the *Badminton* was launched, the funnel colours on the *Dunkerton* were changed from black with a white band and funnel disc to cream with a white band and funnel disc. After that, all of Bristol Sand & Gravel's ships had the same colours.

When built in 1950, *Camerton* was the largest sand dredger working on the Bristol Channel, and for the first six months, because she was the flagship of the fleet, her hull was painted cream. At the time, Frederick Peters treated her as his personal yacht, entertaining friends and business acquaintances to fishing and day trips.

It was not unusual for British Dredging's ships to ballast their sand tanks with water, in readiness for loading. This was carried out by the shore crane at Pooles Wharf, lifting water from the dock in its grab and emptying its contents into the sand tank of the ship. This practice was only carried out in fine weather, or if the dredger was to run the tide to the Firefly Buoy near Portishead.

British Dredging had a unique company house flag and funnel board. It looked like a rectangle divided into triangles of yellow and blue, with lettering

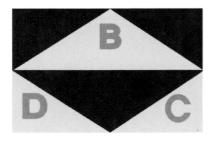

British Dredging Company Ltd logo.

in red. The top-centre yellow triangle and the blue triangles to the sides represented the sand dredged against the sky. The lower blue centre triangle with the yellow triangles on each side represented the sand dredged from the seabed. The lettering BDC was in red on the yellow sections.

The largest of the British Dredging ships to berth at Pooles Wharf, Bristol was the *Bowtrader*, 1,592 gross tons, which was built along the lines of a 'Thames flat iron' for service in the upper Thames. She arrived from her builders in 1969, to be shown to the trade before taking up station on the Thames.

Both the *Camerton* and the *Peterston* were built to the maximum length allowable for Bristol Sand & Gravel's Dundas Wharf, just below Bristol Bridge and opposite the Welsh Back. When fully loaded, neither ship could get right alongside the berth, due to the fact that they touched the bottom of the dock. They had to lie ten feet from the berth until they had discharged 150-200 tons of cargo.

When required, the Bristol sand dredgers had a trade in Avonmouth, ballasting the ships sailing to Australia and Canada.

The ballasting was done over the side using the derrick and winch of the ocean-going ship. The dredger was positioned alongside the ship and her cargo was grabbed from her sand tank using a dumper grab. The loaded grab was lowered into the hold of the ship being ballasted, where dockers released the dumper grab with a release rope. Ballasting could take between twelve and thirty-six hours and may have involved the dredger loading a second cargo, or involving a second dredger. A tally clerk was employed to count the number of grab loads lifted from the dredger.

As it would be sold at the ship's port of arrival, the dredger's master had to fill out a long form of certification about the sand. The main requirement was that it was clean sand and washed with water to make it low in salt content.

Many years ago, my father told me a true story about two anchors. He had loaded a cargo at the East One Fathom Buoy when the weather suddenly changed, so he ran for shelter in Cardiff Roads and anchored. On picking up the anchor to proceed into Cardiff, the crew had trouble breaking it free from the seabed. As the anchor broke the surface, it was found to have fouled an anchor lost by another ship. After some time and a struggle, this anchor and a length of chain were brought on board; examination showed it to be of Dutch origin, probably lost by a Dutch coaster. This anchor remained on the ship's deck and was later unloaded in Bristol. Some months later, the ship lost her starboard anchor and the Dutch anchor was fitted as a temporary replacement.

Unloading at Dundas Wharf, Bristol.

The *Bowline* is taking on water from the crane grab at Pooles Wharf.

This is not the end of the story. The following winter, anchored in Cardiff roads, the Dutch anchor fouled on the seabed and it took great effort, using the ship's engine and windlass, to break it free. On breaking the surface, my father discovered that it had fouled the original anchor. This anchor was secured and landed at Cardiff using the shore crane. The starboard anchor was later refitted and the Dutch anchor was kept as a spare. Unbelievable as this sounds, it is a true story.

A weather-related incident occurred on 9 February 1963, when the Bristol Sand & Gravel ship *Badminton* was bound to Bridgwater loaded with 600 tons of gravel.

The River Parrett was covered in ice due to severe winter weather. As the *Badminton* proceeded upriver, the ice became steadily thicker until it became pack ice. There were also biting gale-force winds. The ship was caught in a sudden gust, which turned her ninety degrees, causing her to ground at a right angle to the river bank. As the tide dropped, the *Badminton* was left grounded on pack ice with the worry that she might break her back. Fortunately, she refloated on the next tide, proceeded to the Bridgwater wharf, discharged, and returned to Bristol for inspection. This inspection showed severe bottom damage, and she was sent to Ailsa Shipbuilding in Troon for repair.

A story goes that the *Alwin*, formerly the *Rapid*, was loaded and was waiting to enter the lock at Hotwells. While she was waiting, the tide started to ebb and the lockmaster hailed the *Alwin* to come into the lock. Although her captain drove her ahead at full speed, she only just managed to stem the outgoing tide and enter the lock, avoiding another several hours wait and a possible grounding.

Bunkering on the motor ships and the oil-fired steamship *Camerton* was a relatively easy task carried out by a road tanker pumping the fuel oil into the bunker tank of the ship. However, on coal-fired steamships like the *Dunkerton*, it was a very noisy and extremely dusty operation. Coal was brought to the quayside at Cardiff in railway wagons; the shore crane then grabbed the coal from the wagons and tipped it into the bunker hold on the ship. On deck, it was noisy, and clouds of black dust came from the bunker hold to descend on the whole ship. Down below, throughout the entire operation, it was thunderous each time the grab dropped its load down the bunker hold. On completion, every part of the ship's deck had to be hosed down to remove sometimes as much as an inch of coal dust.

The railway wagons were moved along the quay in two ways, either by using the cranes grab in the wagon, or, more commonly, by using the ship's windlass and capstan to move the wagons forward and under the crane.

Each of the wagons had a ticket under a clip on the chassis side. The ship's engineer removed these tickets and filed them as proof of how much coal was loaded and from which mine it had come.

Notes on Pilotage

The captains and mates of the Bristol Channel aggregate dredgers were expected, by their ship's owners, to undertake their own pilotage in the Bristol Channel and within their ports of call. The only time a pilot would be employed was at ports such as Bridgwater, Watchet, and in the River Severn, where pilotage was compulsory for ships and masters not based at these ports. Pilotage was also compulsory when the ship was over a certain length or tonnage, or not a regular visitor to that port. A master could hold an exemption certificate, which was normally issued after several visits supervised by a qualified pilot. Ships over a certain length would also be required to employ a tug in the River Avon and River Severn.

New vessels leaving Charles Hill's shipbuilding yard on sea trials were required by their insurance to have both a qualified pilot and tug in attendance, and throughout the sea trials they would fly the house flag of Charles Hill & Sons. Although the ship was painted in her intended owner's colours, she remained the property of her builders until handed over after successful sea trials, and until her handover, she was insured by her builders.

Another time a pilot and tug would be employed was when a new ship such as the *Peterston* arrived for the first time on her maiden voyage. The insurance requirement stated the employment of both for safety reasons.

Normally, the tug did no work but simply escorted the vessel upriver; both pilot and tug were there to assist should the ship's master require them. For example, aggregate dredgers from London working for British Dredging would employ a local master to pilot the ship and, if required, hire a port tug. At Bristol, a local master, a pilot, and a tug would be employed, because of dangers in the River Avon.

It should be remembered that the Bristol Channel is unique. It has the second highest rise and fall of tide in the world, second only to the Bay of Fundy in Canada. The rise and fall is forty-seven feet, with a flow in excess of five knots. This is in calm conditions; in bad weather, the in-tide can make an extra two or three feet. For experienced local crews, this presents possible difficulties: for crews who are strangers to the channel, it could easily become overwhelming.

Employment of pilots and tugs is done only when necessary, as this adds to the cost. Wherever possible, the crews are trained to pilot their vessels at sea and in and out of berth. The governing factor is always defined in two ways: insurance requirements, and local port regulations.

Docking at Bridgwater Dock and Bathurst Basin

The Holms Sand & Gravel ships trading to Bridgwater Dock had compulsory pilotage, both in the river and berthing in the dock itself. It was necessary to take a pilot under local regulations. The river's navigable channel was constantly altering and the Bridgwater pilots would regularly walk the river to note any changes that had taken place. Docking at Bridgwater was a difficult operation, and a ship like the *Steepholm* was designed and built to the maximum size for use at the port.

On arrival outside the entrance to Bridgwater Dock, and also Bathurst Basin, the ship was required to turn ninety degrees to enter the lock from the river. This was achieved with careful use of steering and engine power, together with ropes to warp the ship into the lock. Warping was done by moving the ship forward slowly: running the forward spring (a rope) up the lock, one bollard at a time, while effectively using both the engine power and steering. Nowadays, this manoeuvre would be achieved more quickly and easily with the use of a bow thruster to turn the bow of the ship.

Discharging at Bathurst Basin was always difficult, especially at night when it was important not to disturb the patients at the nearby Bristol General Hospital. Only pump discharge was permissable during night hours to reduce noise.

Acknowledgements and Sources

I owe a great debt of gratitude to my very good friends Brian and Margaret Lewis: to Brian for his technical help with the photography and the layout of the book, and to Margaret for advice and for initially editing my notes. There is no doubt that without their help there would be no book. It is good to finally see the research of a lifetime brought together as a publication.

I am also most grateful to Cedric Catt for the use of his photographic archive, and also to the late Jim Crissup who taught me so much and left a splendid maritime archive, now sadly dispersed.

I am indebted to Geoff Williams for his time spent checking the technical content of this book.

I must also thank Amberley Publishing for the opportunity to turn my notes and records into published form, and to record a most important part of our maritime history.

INFORMATION
Appledore Shipbuilders Ltd, ARC Marine Ltd, Dr Andrew Bellamy, Bristol Central Library, Bristol City Council, Bristol Evening Post, Bristol Metal Spraying & Welding Co. Ltd, British Dredging Co. Ltd/British Dredging Aggregates Ltd, British Marine Aggregate Producers Association, Captain John Campbell, Cedric Catt, Cemex Ltd, Charles Hill & Sons, Arthur Clark, Crown Estate, The Friends of the Newport Ship, The Geological Society, The Golden Harvest – R. Parsons & P. R. Gosson, The Gosson Family Archive, William Hamlin, Peter Harman, Captain Peter Herbert, Historic Environment Record (Peter Insole), Dave Hunt, Jim Crissup Archive, Toby Jones of Newport Museum, Brian Lewis, Lloyds Register, Captain Danny Lynch, Marine Aggregate Dredging, Marine Aggregates, Miramar Ship, Graham Mobbs, Bill Moore, Peter Naish, The Port of Bristol Authority, Ready Mix Concrete Ltd, Ron Richardsen, Captain Chris Reynolds, Philip Rudd, Vernon Stone, Wessex Archaeology, Geoff Williamslty.

PHOTOGRAPHY
Paul Barnett. Bristol Sand & Gravel Co. Ltd, British Dredging Co. Ltd, The Cedric Catt Collection, Arthur Clark, John Clarkson, Richard Cornish, Peter Glenn, Peter Gosson and The Gosson Family Archive, William Hamlin, Peter Harman, The Jim Crissup Archive, Brian Lewis, Captain Danny Lynch, Graham Mobbs, Bill Moore, Andrew Palmer, Reynolds Collection, Ron Richardsen, Risdon Beazley Ltd, Philip Rudd, South Coast Shipping Ltd , South Wales Sand & Gravel Co. Ltd (John Bevan), Vernon Stone, Geoff Williams

If I have overlooked anyone or inadvertently infringed copyright claims please accept my honest apologies.

ALSO AVAILABLE FROM AMBERLEY PUBLISHING

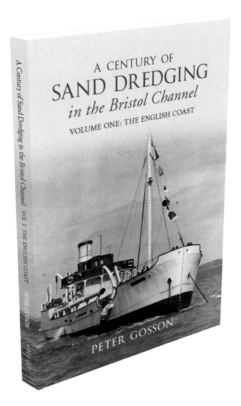

A Century of Sand Dredging in the Bristol Channel
Volume One: The English Coast
Peter Gosson

The first half of a detailed history of sand dredging in the Bristol Channel.
Volume One: The English Coast examines the history of the English sand trade,
the sand dredgers employed through the years, and the English companies
engaged in the trade.

235 x 156 mm | paperback | 160 pages | 195 black & white illustrations
ISBN: 978-1-84868-791-2 | £16.99